Michael Röhrig

Fabrication and Analysis of Bio-Inspired Smart Surfaces

Schriften des Instituts für Mikrostrukturtechnik
am Karlsruher Institut für Technologie (KIT)
Band 22

Hrsg. Institut für Mikrostrukturtechnik

Eine Übersicht über alle bisher in dieser Schriftenreihe
erschienenen Bände finden Sie am Ende des Buchs.

Fabrication and Analysis of Bio-Inspired Smart Surfaces

by
Michael Röhrig

KIT Scientific Publishing

Dissertation, Karlsruher Institut für Technologie (KIT)
Fakultät für Maschinenbau

Tag der mündlichen Prüfung: 17. Dezember 2013
Hauptreferent: Prof. Dr. Volker Saile
Korreferenten: PD Dr. Hendrik Hölscher, Prof. Dr. Metin Sitti

Impressum

KIT Scientific Publishing

Karlsruher Institut für Technologie (KIT)
KIT Scientific Publishing
Straße am Forum 2
D-76131 Karlsruhe

KIT Scientific Publishing is a registered trademark of Karlsruhe Institute of Technology. Reprint using the book cover is not allowed.

www.ksp.kit.edu

Print on Demand 2014

ISSN 1869-5183
ISBN 978-3-7315-0163-3

Karlsruhe Institute of Technology

Fabrication and Analysis of Bio-Inspired Smart Surfaces

Zur Erlangung des akademischen Grades
Doktor der Ingenieurwissenschaften
von der Fakultät für Maschinenbau des
Karlsruher Instituts für Technologie (KIT)

genehmigte
Dissertation
von

Dipl.-Ing. Michael Röhrig

Tag der mündlichen Prüfung: 17.12.2013

Hauptreferent: Prof. Dr. Volker Saile
Korreferenten: PD Dr. Hendrik Hölscher
Prof. Dr. Metin Sitti

Preface

This work would not have been possible without the inspiring discussions that I had with a large number of colleagues and the support of many people and institutions. Below, I would like to single out a few for special recognition.

First of all, I would like to thank Prof. Dr. Volker Saile, for giving me the opportunity of performing research at the *Institute of Microstructure Technology (IMT)* at the *Karlsruhe Institute of Technology (KIT)*. I wish to express my gratitude to him for ensuring optimal working conditions, and stimulating discussions. Furthermore, it is a pleasure to acknowledge Prof. Dr. Jürg Leuthold for his great support and his interest in the upscaling of micro- and nanoreplication technologies.

Particularly, I am very much indebted to my supervisor, PD Dr. Hendrik Hölscher, for his excellent support throughout my thesis. During the years of trustworthy cooperation he impressed me in both professional and interpersonal regard. During that period of time, I came to learn many things from him and became acquainted with the physicist's perception of science and technology. It was a great pleasure to discuss and attack the exciting challenges together.

At this point, I would like to acknowledge the *Nanoscribe GmbH* for providing the 3D direct laser writing system used in this work. By name I would like to thank Martin Hermatschweiler, Alexander Legant and, in particular, Dr. Michael Thiel. Without their support, rapid prototyping of gecko-inspired adhesives would not have been successfully.

I am very much indebted to Prof. Dr. Metin Sitti, Professor for Mechanical Engineering at the *Carnegie Mellon University (CMU)* with emphasis on micro- and nanoscale robotics, for inviting me to join his group in autumn/winter 2011. It was a great pleasure to work with him, and exciting to be a part of this highly innovative group. In particular, I am indebted to Yiğit Mengüç for providing the testing setup for the self-cleaning measurements, his help in data acquisition and analysis and his friendly support outside the laboratory. Furthermore, thank you to Uyiosa Abusomwan who basically set up the modeling of the self-cleaning mechanisms. I'd like to thank Jiho Song for training me in soft-molding and James Weaver for taking SEM images. Furthermore, I would like to thank Burak Aksak from *nanoGriptech LLC* for providing the elastomeric microfibers. Additionally thank you to Shuhei Miyashita, Onur Ozcan, Eric Diller and the other group members for their pleasant hospitality. For funding my research period at the Carnegie Mellon University (CMU), I would like to thank the *Karlsruhe House of Young Scientists (KHYS)*.

It is a great pleasure to acknowledge my colleagues at *IMT* for the good cooperation and pleasant working atmosphere. I wish to express my gratitude to PD Dr. Matthias Worgull, Markus Heilig and Dr. Alexander Kolew for inducting me into the secrets of polymer micro- and nanoreplication. In particular, thank you to Marc Schneider who was always on hand with help and advice. This thesis would not have been possible unless their amicable support and the day and night access to the hot embossing machines. In this regard, I would like to acknowledge the support of the *Karlsruhe Nano Micro Facility (KNMF)*, a Helmholtz Research Infrastructure at Karlsruhe Institute of Technology (KIT).

Furthermore, I am very much indebted to the students I supervised. It was a pleasure to work with Farid Oulhadj and Guénola Etienne who assisted me in the development of hierarchical hot embossing and hot pulling. Additionally, I would like to thank Gerlinde Utsch and Senta Schauer for supporting the replication of hierarchical structures and the nanofur. A thank you to Moritz Baier for his assistance in preparing the journal cover for "Small". I would like to express my gratitude also to

Henri Louvin, Annika Biro, and Andreas Hopf for their contribution to the characterization of the nanofur.

Most of the SEM images shown in this work were captured by Paul Abaffy, whom I would like to thank for the numerous hours in front of the scanning electron microscope. In addition, special thanks to the workshop and clean-room team. To Matthias Mail, thank you for supporting the investigation of the air-retaining capability of the nanofur. In addition, I would like to thank Dr. Stefan Gärtner and Dr. Herrade Bieber who wrote out the patent applications. Particularly, I would like to thank my office mates Taleieh Rajabi, Christian Lay, and Tobias Meier who made me happy to go to work each morning. Thank you to Christian Lay for being my personal LaTeX debugger. I enjoyed the many stimulating discussion, lessons in AFM techniques and electronics given by Tobias Meier - thank you to him for his friendly support.

At KIT it is a pleasure to acknowledge fruitful discussion with Prof. Dr. Juerg Leuthold, Sven Bundschuh, Nicole Steidle, Zhenhao Zhang, Radwanul Siddique, Dr. Maryna Kavalenka, Dr. Julia Syurik, Richard Thelen, Dr. Klaus Feit, Dr. Michael Bruns, Jakob Striffler, Bastian Münster, Christof Megnin, Dr. Thomas Gietzelt, Prof. Dr. Roland Dittmeyer, and Markus Moosmann. Outside KIT I would like to thank Dr. Elmar Kroner (Leibniz Institute for New Materials, Saarbrücken), Dr. Dadhichi Paretkar (Leibniz Institute for New Materials, Saarbrücken), Michael Bartlett (University of Massachusetts, Amherst), Andrew Gillies (University of California, Berkeley), Prof. Dr. Marlene Kamperman (Wageningen University), and Prof. Dr. Stanislav Gorb (Christian-Albrechts-University, Kiel) for inspiring discussions.

Finally, I would like to express my gratitude to my beloved family, especially to my parents and Bianca, for their understanding and loving support.

Karlsruhe, May 2013 *Michael Röhrig*

Abstract

Inspired by nature, numerous innovations such as hook-and-loop fasteners and self-cleaning house paint have entered everyday life in recent decades. Many of these "patents by nature" are closely related to micro- and nanostructured surfaces. Today, bio-inspired products containing the word "nano" are on everyone's lips. However, plenty of stunning properties found in flora and fauna require a complex hierarchical formation of micro- and nanostructures that is not achievable by established polymer replication techniques up to now.

By means of gecko-inspired adhesives, this work introduces novel techniques for the fabrication of hierarchical micro- and nanostructures. After discussing the elementary design principles, a design study of gecko-inspired adhesives fabricated by the flexible 3D direct laser writing technique is shown. The subsequent adhesion analysis performed by atomic force microscopy with colloidal probes reveals the manner in which design affects dry adhesion. Applying soft molding and dipping processes, different sizes of soft mushroom-shaped microfibers were created. By contaminating and cleaning these samples the mechanics of contact self-cleaning were investigated. Exploiting these observations a synthetic gecko-like adhesive was achieved, matching the adhesion and self-cleaning of geckos very closely. To present a cost-effective replication method, the established hot embossing technique was enhanced in this work. By using these advanced hot embossing techniques, gecko-inspired micro- and nanostructures that possess three levels of hierarchy were fabricated. Slightly changing these processes allows for the creation of high aspect ratio nanofur which is superhydrophobic, superoleophilic, underwater air-retaining, and even self-healing when surface treated.

By introducing very flexible, as well as scalable replication techniques, this work offers fabrication solutions for most of the demands on bio-inspired surfaces existing in research and industry.

Kurzfassung

Inspiriert durch die Natur sind in den vergangenen Jahrzehnten eine Vielzahl technischer Innovationen, wie beispielsweise Klettverschlüsse oder selbstreinigende Fassadenfarbe, in unser tägliches Leben getreten. Ein Großteil dieser "Patente der Natur" steht in enger Beziehung zu mikro- und nanostrukturierten Oberflächen, wobei eine Vielzahl der angestrebten Oberflächeneigenschaften eine sehr komplexe Ausgestaltung filigraner hierarchischer Strukturen erfordert. Durch gängige Replikationsverfahren können diese jedoch häufig nicht verwirklicht werden.

Am Beispiel von Adhäsiven nach dem Vorbild der Gecko-Haftzehen werden in dieser Arbeit neuartige Verfahren zur Herstellung hierarchischer Mikro- und Nanostrukturen eingeführt. Auf die Diskussion der grundlegenden Designprinzipien folgt die Analyse gecko-artiger Adhäsive, welche durch direktes Laserschreiben hergestellt wurden. Die darauffolgende rasterkraftmikroskopische Untersuchung der Haftkräfte mithilfe von Kolloid-Sensoren zeigt auf, in welcher Art und Weise die erzielbare Adhäsion durch das Strukturdesign beeinflusst wird. Mithilfe einer Kombination aus Gieß- und Eintauchprozessen wurden pilzförmige Mikrostrukturen verschiedener Größe hergestellt. Durch die Verschmutzung und anschließende Reinigung dieser Strukturen konnte die Funktionsweise der Scher-Selbstreinigung untersucht werden. Auf der Basis dieser Beobachtungen wurde ein synthetisches Klebeband geschaffen, welches sowohl in seiner Haftkraft als auch in seiner Fähigkeit zur Selbstreinigung nahezu dem Gecko-Vorbild entspricht. Um neben den zuvor erwähnten Techniken einen kostengünstigen und skalierbaren Herstellungsprozess aufzuzeigen, wurden neuartige Heißprägetechniken entwickelt. Mithilfe dieser Neuentwicklungen wurde eine mikro- und nanostrukturierte Dreifachhierarchie nach dem Vorbild der Gecko-Haftzehen hergestellt. Durch eine geringfügige Abwandlung der entwickelten Heißprägeprozesse lässt sich ein sogenannter Nanopelz, bestehend aus kleinsten Härchen mit höchsten Aspektverhältnissen, herstellen. Dieser Nanopelz ist superhydrophob, superlipophil, lufthaltend und nach einer entsprechenden Oberflächenbehandlung auch selbstheilend.

Mit der Einführung äußerst flexibler auf der einen, sowie skalierbarer Replikationstechniken auf der anderen Seite, eröffnet diese Arbeit Lösungen um nahezu allen Anforderungen gerecht zu werden, die in Forschung und Industrie an die Herstellung biomimetischer Oberflächen gestellt werden.

Contents

1. Introduction

Inspired by the enormous information density of DNA, in 1959 Richard P. Feynman laid the foundation of modern micro- and nanotechnology in his visionary talk *"There's Plenty of Room at the Bottom"*.[1] The technological progress in micro- and nanostructuring that has occurred over the past 50 years nowadays enables the fabrication of smallest transistors[2] that provide sufficient processing power to compute face recognition, a feature Feynman was dreaming of.[1] Novel techniques such as combinatorial synthesis of peptide arrays with a laser printer[3] bring technology a step closer to marvelous biological systems.

One year after Feynman's talk, the scientific field called *bionics* was independently launched by the congress "Bionics Symposium: Living Prototypes - the key to new technology".[4] Presumably, the term *bionics* was derived from the words *biology* and *technics* by the congress participants.[4] Later, the term *bionics* was replaced by *biomimicry* in scientific vocabulary.[5] *Biomimicry* is a composition of the greek words *bios* and *mimesis* which mean *life* and *to imitate*, respectively. In this discipline researchers with engineering, natural science and medical backgrounds collaborate pursuing the idea of Feynman. Throughout millions of years of evolution nature has developed numerous impressive solutions that nowadays inspire scientists to cope with current technological challenges. Biomimicry systematically transfers nature's solutions to technical innovations.[6] These "patents of nature" result in products and applications that deeply affect our everyday life. The popular hook-and-loop fastener, for instance, was inspired by burdock burs that are very hard to remove from clothing. The fundamental patents of hook-and-loop fasteners still achieve license fees of billions of dollars.[7] Certainly, many more of these "innovations inspired by nature" will follow in the future.

1.1. Motivation

Plenty of stunning properties found in flora and fauna are closely related to micro- and nanostructures. The famous self-cleaning effect of the *sacred lotus*, for example, is based on the surface composition of their leaves.[8] They possess micron sized papillae that are covered with hydrophobic epicuticular wax crystals.[8] Due to this surface topography, contaminants which are generally larger than the micro- and nanostructures come in very little contact with the leaves.[9] Therefore, water that rolls over the water-repellent leaves easily captures and removes these contaminants from the surface.[9]

Another amazing example of evolution is the air-retaining capability of the *salvinia molesta*.[10] The leaves of this floating water fern are covered by a hierarchical formation of multicellular hairs covered with wax crystals that make them hydrophobic. These hairs eventually join at their terminal end and form a flat patch. In contrast to the hairs, these patches are smooth and hydrophilic.[10,11] When submerged underwater, the hydrophobic wax crystals prevent water from penetrating between the hairs, whereas the hydrophilic patches are pinning the water. Therefore, salvinia leaves are able to maintain a robust air layer between their hairs underwater.[10] In such ways, the leaves of the *sacred lotus* and the water fern *salvinia* ensure unobstructed photosynthesis even under extreme conditions.

Besides flora, self-cleaning is also observed in fauna, even though this is due to another motive. *Geckos* for example exhibit smart cleaning of their adhesive pads while attaching and detaching their toes from the substrate.[12,13] By this means, geckos can maintain sufficient adhesion in dirty habitats. In fact, geckos have an enormous attachment strength of around

10 N per 100 mm^2 of pad area (*Gekko gecko*).[14] One of the biggest geckos, a tokay gecko (Figure 1.1), is even able to cling to a surface with a single toe.[12] The high adhesion is attributed to the micro- and nanobristles covering the toe pads of these geckos. Due to their hierarchical design, the adhesive pads are very supple and allow very intimate contact even with rough substrates. The high contact area that geckos achieve with the substrate enables them to easily climb nearly any surface mainly with the help of van-der-Waals interactions.[15]

These bio-inspired effects are of interest not only for fundamental research[18–22] but also for technical applications such as self-cleaning antennas,[23] anti-fouling (Figure 1.2) and drag reducing coatings,[10] climbing robots[24,25] and industrial robotic pick-and-place manipulators,.[26,27] From 2005 until 2008 the hundred most prosperous biomimicry-based products including architectural projects generated more than \$1.5 billion in revenues.[28] In 2009, the U.S. Patent and Trademark Office (USPTO) received more than 900 patents containing the word "biomimicry".[29] Consequently, this field is very attractive for investors and companies as they can attain

Figure 1.1.: Photograph of a tokay gecko (*Gekko gecko*).[16] Since this species is one of the biggest geckos tokay geckos have elaborate attachment pads, making them the best studied geckos.[17]

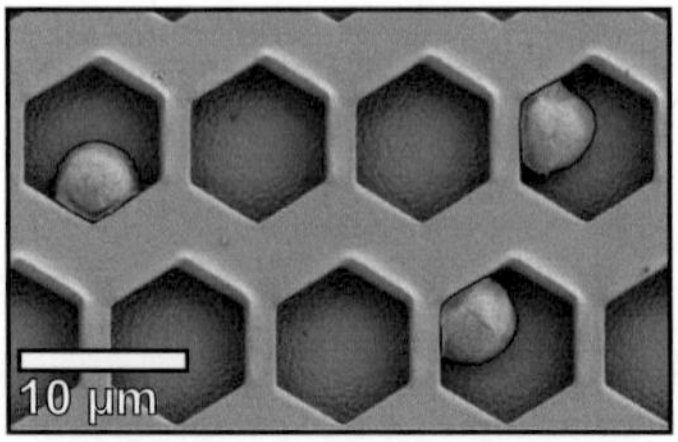

Figure 1.2.: Scanning electron microscopy image showing the settlement of zoospores of the green alga *Ulva linza* on hot embossed ‘honeycomb’ gradient microstructures. Interestingly, ‘kink sites’ of the microstructure resembled preferred attachment positions for the zoospores. The zoospores didn’t settle on pits that were smaller than 2.6 μm indicating that zoospore settlement is remarkably reduced on smaller micro- and nanostructures.[30]

returns of 40 - 50 % on new bio-inspired products.[29] Since companies all over the world are active in the field of biomimicry, this sector is estimated to represent about $1.0 trillion of the global gross domestic product in 2025.[29]

Consequently, the fabrication of novel bio-inspired engineered surfaces is pursued by industries and scientists all over the world.[31–36] For example, inspired by the *sacred lotus* synthetic self-cleaning surfaces that are extremely water-repellent were fabricated in this work. Due to the enormous superhydrophobicity of the microstructured PTFE surface, water droplets even bounce on these surfaces (Figure 1.3).

In order to fabricate bio-inspired smart surfaces, typically, the established processes for the replication of micro- and nanostructures are applied. Well established in industries, *injection molding*[37] is the most common technique for mass fabrication. In recent years, *UV-nanoimprint* increasingly gained significance due to its high precision.[38] *Hot embossing* is the most universal micro-replication technique, which makes it particularly attractive for research facilities.[39] However, the complex micro- and nanostructures found in nature usually require advanced fabrication techniques. The limitations in achievable shapes, aspect ratios, multilevel hierarchy and scalability

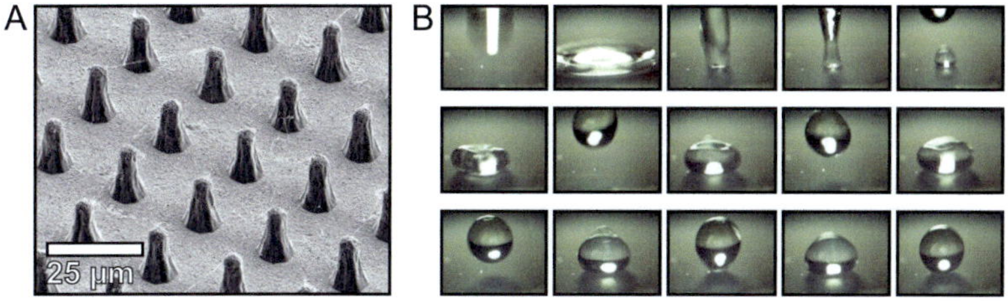

Figure 1.3.: Superhydrophobic microstructured surface inspired by the *sacred lotus*. **A** Applying hot embossing, a PTFE surface covered by micro pillars was created (∅: 5 - 12 µm, height: 21 µm, period: 22 µm). **B** Due to the low surface energy microstructures, droplets applied to the superhydrophobic surface even start to bounce.

of these technologies impede commercial break-through of various bio-inspired smart surfaces.[33,39–43]

For most technical applications it is not advisable to copy nature's micro- and nanostructures 1:1.[4] In fact, they have to be adopted to the available materials and processes. This requires an in-depth understanding of the biological, chemical and physical correlations the discovered effect is based on. For fundamental research, however, extremely flexible fabrication techniques are required to study these correlations experimentally. Established replication techniques are not suitable for this purpose, since they impede investment costs for the fabrication of mold inserts.

1.2. Objective and Outline

In 1959 Feynman already raised the questions *"What kind of manufacturing processes would we use? [...] What are the limitations as to how small a thing has to be before you can no longer mold it?"*.[1] By means of gecko-inspired adhesives, this work introduces novel micro- and nanostructuring techniques and advanced molding processes for the cost-effective replication of hierarchical micro- and nanostructures. New insights into robust, self-cleaning adhesives are generated by analyzing the fabricated gecko-inspired structures. By applying the developed replication technologies, superhydrophobic nanofur which mimicks the water repellency of lotus, and air-retaining of salvinia leaves is created.

To summarize the principles of gecko adhesion, I give a brief overview of this inspiring attachment system in Chapter 2. As further discussed in Chapter 2, robust adhesion of gecko-inspired adhesives requires accurate adaptation of the micro- and nanostructures to the chosen material and its properties. In this way, functional failures such as fibrillar bunching can be prevented and adhesion is maximized.

In Chapter 3, I introduce 3D direct laser writing as a rapid prototyping method that yields highest flexibility in the design of gecko-inspired adhesives. Furthermore, I explain the principles of AFM adhesion measurements with colloidal probes. By investigating a design study of gecko-inspired micro- and nanostructures that vary in order of magnitude, amount of hierarchical levels, pitch, aspect ratio and tip shape I reveal how geometry affects their dry adhesion.

Self-cleaning of synthetic gecko-inspired adhesives is the focus of Chapter 4. Applying soft molding and dipping processes, three sizes of mushroom shaped microfibers were created. Investigating their contact self-cleaning ability revealed distinct contact self-cleaning modes. Fortunately, the synthetic microfibers combine high adhesive strength and remarkable adhesive recovery through contact self-cleaning, both comparable to geckos.

In the beginning of chapter 5 I review the established replication processes injection molding, UV-nanoimprint and hot embossing. After defining the challenges of replicating bio-inspired micro- and nanostructures, I present the novel advanced hot embossing processes *hierarchical hot embossing* and *hot pulling*. The enormous potential of these processes was successfully shown in the fabrication of gecko-inspired threefold hierarchical micro- and nanostructures that exhibit remarkable adhesion. Lastly, I discuss the technological limits of these advanced hot embossing techniques.

In Chapter 6, I present a cost-effective hot pulling process for the fabrication of superhydrophobic nanofur that is beneficial for various biomimetic applications. To investigate the spectrum of applications, the wetting behavior of the samples was characterized by measuring the static/dynamic contact angles and the corresponding sliding angles. By damaging the surface structure, liquid traps were created by changing the local wettability. To overcome wearing issues, self-healing was achieved by locking an intermediary liquid acting as water and oil repellent surface. Mimicking the leaves of the floating water fern *Salvinia* the nanofur retains air when submerged underwater. Due to its superhydrophobicity and superoleophilicity, the nanofur is applicable for oil/water separation and oil spill clean up.

In the final chapter, Chapter 7, I summarize the results achieved in this work and reflect on the possible future of bio-inspired smart surfaces. Novel material classes like shape memory polymers and liquid wood offer promising applications such as switchable wettability and a high degree of bioresorbability, respectively.

2. Robust Gecko-Inspired Adhesion

The phenomenal adhesive properties of gecko toes have been extensively investigated for a long time, resulting in the development of novel adhesive tapes.[20–22,44] Besides the strong adhesion to nearly any substrate, their outstanding attachment system allows geckos to detach within milliseconds.[45] This remarkable combination of strong attachment as well as rapid and easy detachment originates from the hierarchical design of delicate hairs covering the lamellae that are crossing the toe-pads (Figure 2.1). These keratinous hairs are called setae and are about 4 μm in diameter and 100 μm in length. Finally, they branch into hundreds of tiny endings, *spatulae*, which are about 200 nm wide each (*Gekko gecko*).[17] Interestingly, the adhesive toes are non-sticky at first, but are activated by a short shear motion that aligns the setae.[17] This default non-sticky state enables geckos to detach easily by peeling their toes off the substrate.[46] Due to the hierarchical design of their attachment system, geckos achieve very intimate contact to flat and even to relatively rough surfaces. The large contact area enables them to climb walls and ceilings only with the help of van-der-Waals interactions.[15] In the presence of humidity, the adhesion is even enhanced.[47]

2.1. Gecko's Secret

In the animal world, hairy attachment systems are commonly utilized by beetles, flies, spiders, and geckos.[49] However, geckos are special. They have a larger body mass than all the other animals that use hairy attachment systems.[50] Moreover, geckos do not secrete sticky glue, as, for example,

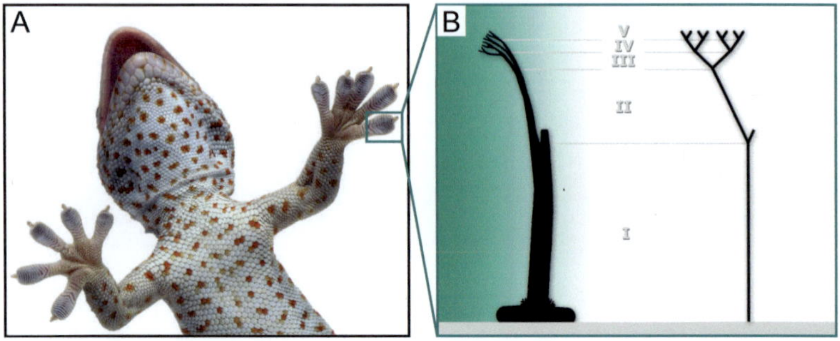

Figure 2.1.: **A** A bottom view of a Tokay gecko (*Gekko gecko*).[16] The toe pad of geckos is typically separated into distinct lamellae. **B** These lamellae are covered with millions of delicate hairs, the so-called setae. The setae are multi-scale hierarchical structures, which are split up to four times. The most tiny endings are the so-called spatulae with a width of about 200 nm (inspired by Batal[48]).

flies do.[17] Instead, they are using dry adhesion based only on van-der-Waals forces.[15]

The significance of micro- and nanoscale fibers and hierarchy in hairy attachment systems was extensively investigated in the past decades. It is known that with increasing body mass of the animal, the efficiency of its hairy attachment system has to be optimized, since the enlargement of the adhesive pads does not linearly scale with their gain in weight.[51] Biological studies revealed two interesting correlations: with an increasing body mass of the animals, the hair density[52] and the amount of hierarchy levels strongly increases.[50] These findings indicate the relevance of both design features.

2.1.1. Fibrillar Attachment System

The evolution of hairy attachment systems is very successful and many animals, with geckos leading the way, benefit from this design.[49] The advantages of this fibrillar design are shortly reviewed below.

Compliance. Since van-der-Waals forces interact over a distance on the order of few nanometers,[53] the adhesion of the attachment system is strongly related to the real contact area achieved with the substrate. Bringing a flat adhesive in contact with a rough substrate requires comparatively high loads in order to achieve intimate contact. While conforming around the substrate's asperities, elastic energy is stored in the adhesive that works against the adhesion process.[50] Decreased rigidity of the adhesive leads to enhanced compliance. Long fibrillar hair can easily bend and enhance compliance drastically.[54]

Contact splitting. Splitting up the contact leads to increased adhesion that is the result of multiple mechanisms. Arzt *et al.*[52] reported that when splitting a contact into n sub-contacts by self-similar scaling, the total pull-off force increases by a factor of $\sqrt{n}$.

Another important advantage is the prevention of crack propagation. Such defects may occur in the presence of contaminants or roughness of the substrate.[52] Since multiple contacts exist in a fibrillar attachment system, a crack that occurs in one of these contacts does not weaken the other bonds.[50] In comparison, in a flat-to-flat contact much less energy is required for a crack to propagate throughout the whole contact.[50] Consequently, the flaw tolerance is drastically increased by fibrillar adhesives.[52]

Optimized stress distribution. Typically, for comparatively large contacts detachment is initiated at the edge of the contact due to the remarkable stress concentration occurring there.[53] Unlike large contacts, tiny spatulae- or mushroom-shaped tips feel uniformly distributed stress.[54,55] In this way the entire interface fails simultaneously and adhesion can be maximized up to the theoretical adhesion strength.[53,54]

Energy dissipation. The fibers pulled off the substrate elongate before detaching.[50] After detachment, the stored energy dissipates and does not contribute to the detaching of neighboring fibers.[54] The energy dissipation is directly proportional to the height of the fibers, making high aspect ratio hair advantageous for the robustness of hairy attachment systems.[53]

2.1.2. Hierarchical Attachment System

It is not a coincidence that the attachment system of tokay geckos exhibits five hierarchy levels. How adhesion benefits from hierarchy is briefly discussed below.

Compliance. Natural surfaces show a wide range of roughness, ranging from microscopic to macroscopic. The hierarchical design of geckos' attachment system with macroscopic lamellae, microscopic setae and nanoscale spatulae enhances the compliance to these roughness profiles by ensuring a low effective elastic modulus.[50] In fact, Autumn *et al.*[17] measured an effective elastic modulus of about 100 kilopascals in isolated arrays of keratinous setae of tokay geckos, despite the rigidity of bulk β-keratine that ranges from 1 to 4 gigapascals.[45,56,57]

Robustness. Compliance and therefore adhesion can be enhanced by increasing the aspect ratio of the fibers or by making them softer. However, doing so eventually results in functional failure by collapse or conglutination.[50] A hierarchical design, however, achieves compliance while preventing fiber bunching.[50] Furthermore, all hierarchy levels contribute to energy dissipation by dissipating elastic energy that is stored during pull-off. In this way, the required work of adhesion and therefore robustness is increased.[58]

2.2. Robust Design

Increasing the aspect ratio of fibers or making them softer is beneficial for compliance, however, fiber collapse eventually occurs.[50] In order to ensure robust adhesion, the design of gecko-like adhesives requires immense optimization to prevent such functional failure of the adhesive tape such as fibrillar bunching. Below, the existing models of designing gecko-inspired adhesives are discussed, expanded and modified for this work.[53,54,58–72]

At first, an adhesive structure consisting of fibers with radius R, length L and inter-fiber distance $2D$ is considered. Assuming further that the Johnson-Kendall-Roberts (JKR)[73] theory can be applied, the pull-off force required to detach a spherical tip from a flat and infinite stiff substrate is given by

$$F_c = \frac{3}{2}\pi R W, \tag{2.1}$$

where W represents the work of adhesion.[73]

2.2.1. Failure Modes

Failure by fiber fracture. In order to increase the adhesion by the principle of contact splitting or enhanced compliance, the fiber radius should be chosen as small as possible.[45,52] However, Spolenak *et al.*[59] proposes that if the fiber radius became too small, the axial stress σ_c may exceed the theoretical fracture strength σ_f resulting in fiber fracture. Hence, the correlation

$$\sigma_c = \frac{F_c}{\pi R^2} \leq \sigma_f \tag{2.2}$$

limits the useful fiber radius R

$$R \geq = \frac{3W}{2\sigma_f} \approx 15\frac{W}{E_f}, \tag{2.3}$$

where the theoretical fracture strength σ_f is approximated by $E_f/10$, with E_f being the elastic modulus of the fiber.[74]

Failure by exceeding the ideal contact strength. The ideal contact strength σ_{th}, that is transmitted through the actual contact area at the instant of tensile instability, is the upper limitation for the contact strength[59]

$$\sigma_c = \frac{F_c}{\pi r_c^2} \leq \sigma_{th}, \tag{2.4}$$

with r_c being the contact radius at the instance of pull-off. Considering the JKR theory and a rigid contacting surface, r_c can be expressed as

$$r_c = \left(\frac{9\pi W R^2 \left(1 - \nu_f^2\right)}{8E_f} \right)^{\frac{1}{3}}. \tag{2.5}$$

Here, ν_f is the Poisson ratio of the fiber. Combining eqs. (2.4) and (2.5), the second lower limit for the fiber radius R can be extracted[69]

$$R \geq \frac{8s^3 E_f^2}{3\pi^2 W^2 \left(1 - \nu_f^2\right)}. \tag{2.6}$$

The characteristic length of surface interaction s is typically in the range of 2×10^{-10}m.[59]

Failure by crack propagation. Due to surface roughness, the mating within the contact zone of fiber and substrate is never perfect. In order to use fracture mechanics to solve this problem, the resulting defects in the interface can be modeled as cracks.[61] Tensile loading of the fiber may cause unstable crack propagation within the interface, leading to reduced adhesion. Considering Griffith's criterion, the critical load for the unstable propagation of an existing crack of the length $2a$ in an elastic solid is[75]

$$\sigma_g = \sqrt{\frac{16G\gamma}{\pi a(\kappa + 1)}}, \tag{2.7}$$

with G representing the shear modulus, γ being the surface energy and κ describing a coefficient depending on the state of plain stress or strain, respectively. Since this equation is only valid for a crack in an elastic solid, it is adapted for this work to describe a crack in the interface of fiber and substrate. As a crack propagates through a brittle solid, the following energies are contributing: the strain energy U, the surface energy of the crack S and the potential energy of the external forces P. Griffith's criterion supposes, that strain energy is released during crack propagation. For an unstable propagation of the crack, the released strain energy has to overcome the required surface energy, hence, resulting in a reduction of the total potential energy.

Consequently, the Griffith's criterion for a crack to propagate by da calculates to

$$\frac{d}{da}(U+S+P) \leq 0. \tag{2.8}$$

Assuming a model of a disk in a field of uniaxial tensile stress, for a Griffith crack which is $2a$ in length, the sum of strain energy and potential energy is[76]

$$U+P = -\frac{\pi(\kappa+1)}{8G}\sigma_t^2 a^2, \tag{2.9}$$

with σ_t being the tensile stress. The surface energy of a crack in the interface between fiber and substrate is

$$S = 2a\left(\gamma_f+\gamma_s\right), \tag{2.10}$$

where γ_f and γ_s are the surface energies of the fiber and the substrate. Combining eqs. (2.8), (2.9) and (2.10) leads to the instability criterion

$$\frac{d}{da}\left[-\frac{\pi(\kappa+1)}{4G}\sigma_t^2 a^2 + 2a\left(\gamma_f+\gamma_s\right)\right] \leq 0. \tag{2.11}$$

Providing that the energy released by extending the crack by *da* overcomes the required surface energy, unstable crack propagation occurs. Hence, the critical Griffith load for a crack within the interface of an adhesive contact between fiber and substrate is

$$\sigma_{fs} = \sqrt{\frac{E\left(\gamma_f + \gamma_s\right)}{\pi a}}, \tag{2.12}$$

assuming $G = E/(2+2\nu)$ for isotropic materials and assuming a state of plain strain ($\kappa = 3 - 4\nu$). In case of ideal contact strength being below the critical Griffith load, unstable crack propagation is prevented

$$\sigma_{th} \leq \sigma_{fs}. \tag{2.13}$$

Hence, in order to tolerate a defect which is 1/4 R wide and existing in the interface between fiber and substrate, the fiber radius has to fulfill the condition

$$R \leq \frac{4E_f\left(\gamma_f + \gamma_s\right)}{\pi \sigma_{th}^2}. \tag{2.14}$$

Failure by bunching. In a fibrillar array, attractive van-der-Waals forces may cause bunching of neighboring fibers, resulting in a decreased compliance of the structures. For a given distance $2D$ between two fibers, a critical length L exists above which bunching occurs. In the following, this work proposes an anti-bunching condition for fibers from the energetic point of view. For simplicity, in this work the anti-bunching condition is calculated for a quadrangular cross-section of the fibers and a quadrangular pattern.

Considering two bunched fibers touching along the distance $L - l$ (Figure 2.2), the equilibrium is dominated by two contributing energies: the

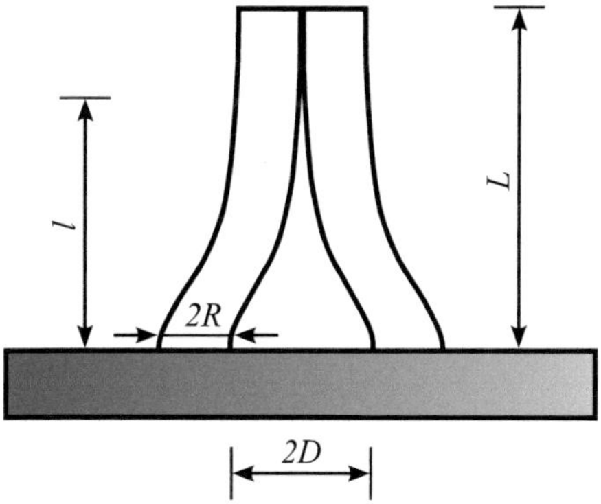

Figure 2.2.: For a given distance $2D$ between two fibers, bunching occurs if the length of the fibers L exceeds a critical value. Van-der-Waals forces forward the bunching, whereas restoring forces of the deformed fibers contribute to the separation of the fibers (inspired by Hui *et al.*[63]).

adhesion along the interface and the potential energy of the deformed fibers. In order to separate two bunched fibers, the energy

$$2U_{adh} = -2 \int_{y=l}^{L} \gamma_f (2R)\, dy = -2\gamma_f (2R)(L-l) \tag{2.15}$$

is required.

In order to model the potential energy of the fibers, a clamped beam of length L is assumed, guided by a floating bearing at l and a linear bearing at its end (Figure 2.3). The bearing force F_b is

$$F_b = 12 \frac{D E_f I}{l^3}, \tag{2.16}$$

with $I = {(2R)^4}/{12}$ being the moment of inertia of a fiber with the quadrangular cross section $(2R)^2$ and D representing the deflection of the fiber.

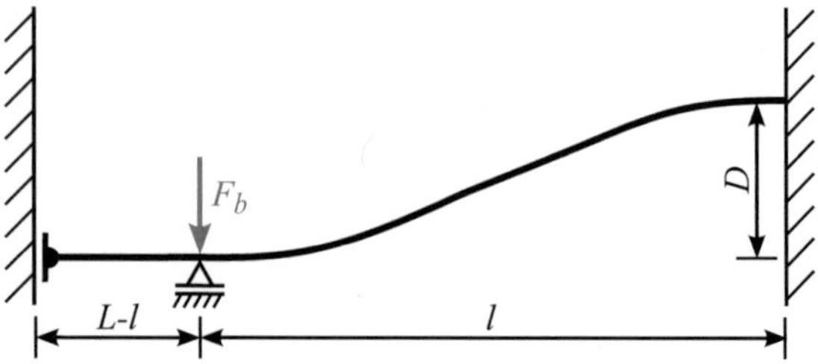

Figure 2.3.: The bunched fiber is modelled as a clamped beam guided by a float bearing at L and a linear bearing at its end.

Hence, the potential energy of the deformed fiber is

$$U_{pot} = \int_{x=0}^{D} F_b x dx = \frac{6E_f I D^2}{l^3}. \tag{2.17}$$

The fibers will tend to adopt the state with the most favorable (i.e. lowest) energy. If no minimum in the effective potential is existing in the interval $0 \leq {}^{l}/_{L} \leq 1$, the fibers do not bunch (Figure 2.4). Summing eqs. (2.15) and (2.17), the effective potential is

$$U_{eff} = U_{adh} + U_{pot} = -\gamma_f (2R)(L-l) + \frac{E_f (2R)^4 D^2}{2l^3}. \tag{2.18}$$

In the following, l/L is substituted by χ for simplicity. In order to prevent a minimum of the effective potential in the interval $0 \leq \chi \leq 1$, the condition

$$U'_{eff}(\chi) = \gamma_f (2R) L - \frac{3E_f (2R)^4 D^2}{2\chi^4 L^3}, \quad 0 \leq \chi \leq 1 \tag{2.19}$$

has to be fulfilled. Setting $\chi = 1$ (critical limit) and solving equation (2.18) defines the maximum fiber length L for bunching prevention

$$L \leq \left(\frac{3}{2}\right)^{\frac{1}{4}} \sqrt{D} \left(\frac{(2R)^3 E_f}{\gamma_f}\right)^{\frac{1}{4}}. \tag{2.20}$$

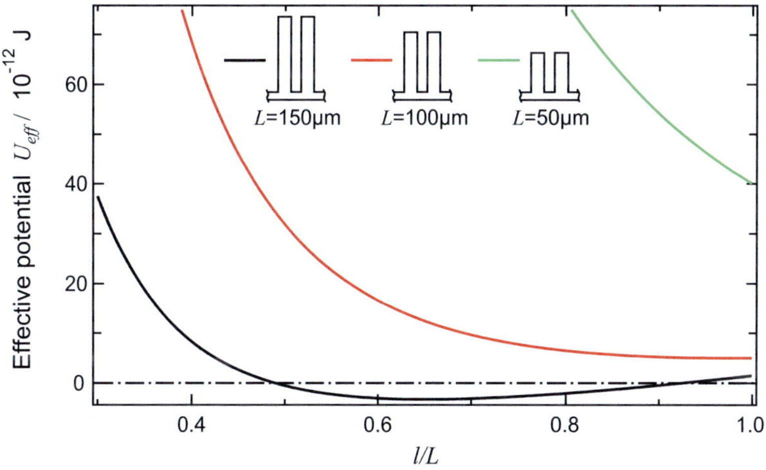

Figure 2.4.: By adjusting the length L of the fibers, a minimum in the effective potential is avoided, preventing bunching of the fibers ($\gamma_f = 33\,\text{mJ}$, $E_f = 4\,\text{GPa}$, $2R = 5\,\mu\text{m}$, $D = 2\,\mu\text{m}$).

The distance between two fibers is $2D$ and can be expressed as a function of the effective area fraction φ and the maximum achievable area fraction φ_{max} for the chosen fibrillar pattern.[61]

$$2D = 2R\left(\sqrt{\frac{\varphi_{max}}{\varphi}} - 1\right), \quad 0 \leq \varphi \leq \varphi_{max} \tag{2.21}$$

Combining equations (2.20) and (2.21) leads to the final anti-bunching condition

$$L \leq \left(\frac{3}{2}\right)^{\frac{1}{4}} \sqrt{R\left(\sqrt{\frac{\varphi_{max}}{\varphi}} - 1\right)} \left(\frac{(2R)^3 E_f}{\gamma_f}\right)^{\frac{1}{4}}. \tag{2.22}$$

2.2.2. Maximizing Work of Adhesion

Considering only van-der-Waals energies and neglecting others, the work of adhesion of a single-level fibrillar array can be expressed as

$$W = \varphi \Delta \gamma = \varphi \left(\gamma_f + \gamma_s - \gamma_{fs} \right), \tag{2.23}$$

where γ_f, γ_s and γ_{fs} represent the surface energies of fiber and substrate, as well as the energy of the fiber-substrate interface. For a multilevel, hierarchical fibrillar array, it is reasonable to add the elastic strain energy to the work of adhesion.[61] Assuming a cylindrical fiber of primary length L, the fiber is elongated by the ideal contact strength by

$$\Delta L = L \frac{\sigma_{th}}{E_f}. \tag{2.24}$$

The work $W_{diss} = \sigma_{th} \Delta L$ required to elongate the fiber is dissipated by inherent material damping when the fiber detaches.[67] In order to calculate the work of adhesion for a hierarchical, e.g. two-level, fibrillar array, this dissipation has to be added to the van-der-Waals contribution

$$W_2 = \varphi_1 \left(\Delta \gamma + W_{diss} \right) = \varphi_1 \left(\gamma_f + \gamma_s - \gamma_{fs} + \frac{L \sigma_{th}^2}{E_f} \right) \tag{2.25}$$

In order to increase the robustness of the adhesive structures, the work of adhesion has to be maximized.

2.2.3. Iterative Dimensioning

According to the previously described limits, gecko-inspired adhesives can be designed robustly following the iterative procedure described in Figure 2.5.

First, the smallest structure level is designed by dimensioning with respect to the laws preventing failure. After defining the lateral dimension, the length and period of the structures can be optimized according to the anti-bunching condition and the maximization of the work of adhesion.

To design the subsequent structure levels, the conditions for the prevention of fiber fracture and exceeding of the ideal contact strength are not required, since the lateral dimension is typically exceeding the previous one. After calculating the lateral dimension in order to prevent crack propagation, length and period of the level are defined considering the anti-bunching condition and the maximization of the work of adhesion.

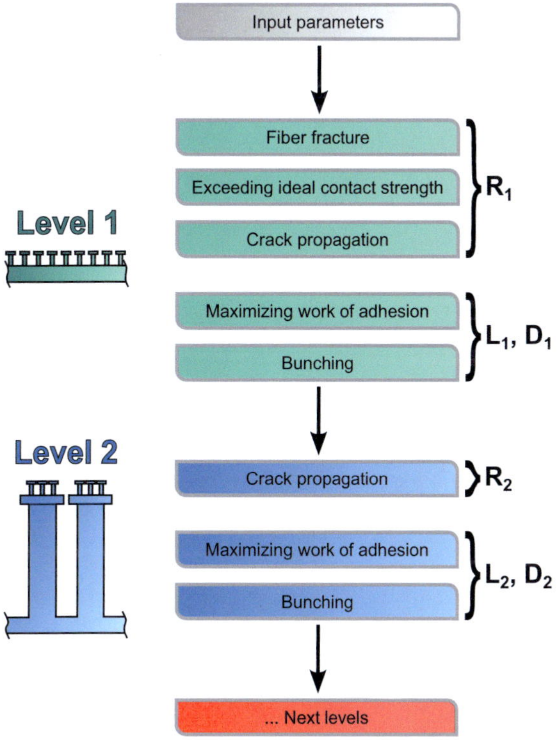

Figure 2.5.: The flow-chart shows the algorithm to calculate the fiber dimensions iteratively for every hierarchy level.

A simulation tool was coded that supports the design of optimized gecko-inspired adhesives (Figure 2.6). After entering the material properties and process restrictions the simulation tool outputs the optimum design for the given boundary conditions. By applying this simulation tool, the gecko-inspired designs presented in this work were optimized.

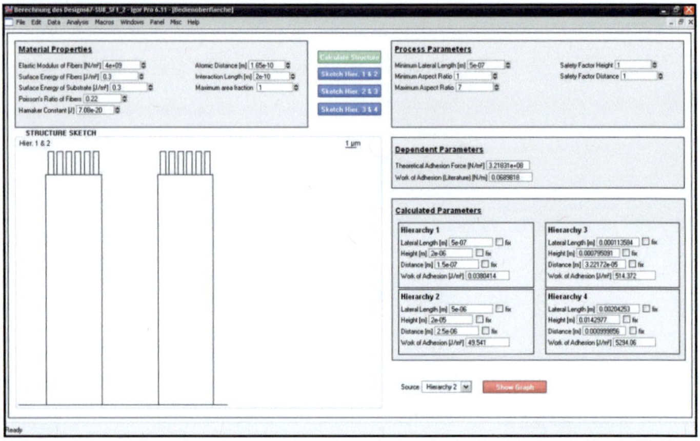

Figure 2.6.: Screenshot of the simulation tool coded in the programming environment *Igor Pro* (Wavemetrics). After entering the material properties and process restrictions, the tool displays the optimized dimensions for the hierarchical gecko-inspired structures.

2.3. Conclusion and Outlook

The enormous adhesion of geckos is based on the micro- and nano-structured hairs covering their toes. The fibrillar and hierarchical design of these setae are the tricky design principles leading to enormous compliance, enhanced adhesion mechanics and a large work of adhesion that altogether ensure robust adhesion.

In designing synthetic gecko-like adhesives, the micro- and nano-structures have to be carefully adopted to the chosen material and fabrication process. In this way, robust adhesion is achievable while minimizing functional failure of the adhesive tape. Considering the physical and mechanical correlations leading to failure, the optimized design can be found by iterative dimensioning. By carefully applying the coded simulation tool in the design phase, fiber bunching and other functional failures that are typical for gecko-like adhesives could soon be a matter of the past.

A shortened version of this chapter was published as the book chapter "How Geometry Affects the Adhesion of Gecko-Like Adhesives". M. Röhrig, M. Thiel, S. Bundschuh, M. Worgull and H. Hölscher: Biological and Biomimetic Adhesives - Challenges and Opportunities, edited by R. Santos, N. Aldred, S. Gorb and P. Flammang. RSC Publishing (2013).

3. 3D Direct Laser Writing of Hierarchical Gecko-Inspired Surfaces

Geckos easily climb walls and even ceilings of nearly any material without leaving any residue behind. Their self-cleaning toes even stick in vacuum or underwater.[12,17] Such properties are economically interesting for climbing robots,[24,25] sports,[77] pick-and-place systems (especially under vacuum),[27,78] as an adhesive on the back of portable devices,[78] and resealable plastic packaging.[79] Consequently, the fabrication of synthetic gecko-like adhesives mimicking these phenomenal micro- and nanostructures is pursued by numerous groups all over the world.[24,32–36,80–87]

Currently, the soft molding technique is the most often used approach for the fabrication of such gecko-inspired adhesives.[32,35,88] Usually, materials applicable for soft molding like PDMS or polyurethane are cast into etched silicon wafers or SU-8 templates. However, the soft molding technique incurs some drawbacks. Due to the demolding process and the need of complex designs for gecko-like adhesives, it is essentially restricted to soft materials. Furthermore, demolding is a delicate process where the mold might be destroyed by accident or sacrificed on purpose.[33,43] Considering that the observed adhesion depends very strongly on structure-design parameters like pillar dimension, aspect ratio and tip shape, mold inserts are an inflexible and exhaustive approach for design studies, because a new mask and/or mold has to be manufactured for every parameter variation.[59,60,89] So far, several geometry induced effects, like the improvement of adhesion by hierarchical structuring, could only be analyzed for soft materials and not for stiff materials at the relevant nanometer scale.[43,81]

In order to surmount these issues, 3D direct laser writing (DLW) is introduced for the rapid prototyping of hierarchical gecko-inspired surfaces with elastic modulus and relevant length scales matching the gecko's toe-pads very closely. In contrast to previous studies normal adhesion experiments were performed on the fabricated gecko-inspired structures. The obtained results show that hierarchical structures are indeed favorable for stiff materials on the nanoscale as expected from numerous theoretical studies.[53,56,60–62,90] In contrast to molding techniques, 3D direct laser writing offers the quick realization of design concepts that are neither restricted to demoldable designs nor limited by any mold fabrication technology. This gives the highest flexibility in creating gecko-mimicking surfaces.

3.1. Fabrication by 3D Direct Laser Writing

Applying 3D direct laser writing, gecko-inspired nano- and micro-structures were fabricated.[91] The dimensions chosen for this study were calculated according to the conditions for robust gecko-inspired adhesion (Chapter 2) and are listed in Table 3.1. The selected acrylic based negative tone resist (IP-G 780) offers highest resolution in multiphoton absorption and enables the design of arbitrary structures down to the nanoscale.[57,92] The elastic modulus of this resist ($E_{IP-G\ 780} \approx 4\,\text{GPa}$) is closer to that of the gecko ($E_{gecko} \approx 1-4$ GPa) than softer materials which are often used for the fabrication of dry adhesives, such as polydimethylsiloxane ($E_{PDMS} \approx 2.6\,\text{MPa}$), polyurethane ($E_{PUR\ ST-1060} \approx 3\,\text{MPa}$), or polyvinyl-siloxane ($E_{PVS} \approx 3\,\text{MPa}$).[43,45,56,57,93,94]

Table 3.1.: Dimensions of all fabricated gecko-mimicking arrays (all values are in µm). The values emphasize the variation of the parameters $h1$ and $p1$ leading to a decreased density (arrays 2, 5, and 8) and aspect ratio (arrays 3, 6, and 9). The values marked with an * correspond to arrays with mushroom-shaped tips.

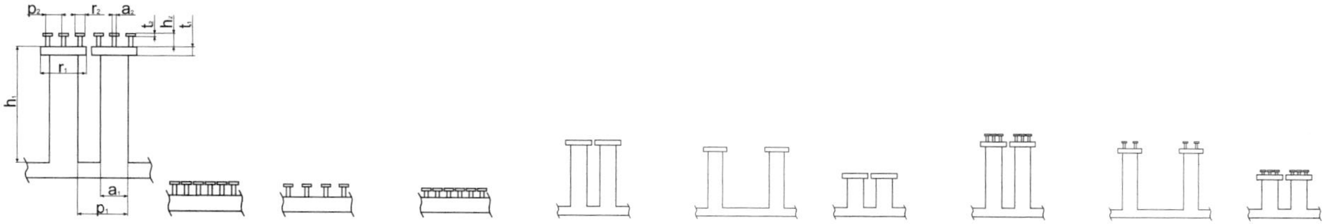

parameter	array number 1	2	3	4	5	6	7	8	9
a_1	0.5/0.56*	0.5/0.56*	0.5/0.57*	5.6/5.6*	5.6/5.6*	5.6/5.6*	5.6/5.7*	5.6/5.6*	5.6/5.6*
h_1	2.25/2.1*	2.25/2.27*	**1.8/1.9***	24.5/27.9*	24.5/27.9*	**16.8/15.9***	24.5/27.5*	24.5/28.4*	**16.8/13.0***
p_1	1.1/1.1*	**1.4/1.4***	1.1/1.1*	9.0/9.0*	**11.0/11.0***	9.0/9.0*	9.0/9.0*	**11.0/11.0***	9.0/9.0*
r_1	0/0.9*	0/0.9*	0/0.9*	0/7.6*	0/7.6*	0/7.6*	0/7.6*	0/7.6*	0/7.6*
t_1	0/0.9*	0/0.9*	0/1.0*	0/2.0*	0/2.1*	0/1.35*	0/2.0*	0/2.1*	0/1.2*
a_2	-	-	-	-	-	-	0.5/0.57*	0.5/0.6*	0.65/0.65*
h_2	-	-	-	-	-	-	1.8/2.7*	2.3/2.7*	1.6/1.0*
p_2	-	-	-	-	-	-	1.1/1.1*	1.4/1.4*	1.1/1.1*
r_2	-	-	-	-	-	-	0/0.9*	0/0.9*	0/0.75*
t_2	-	-	-	-	-	-	0/1.4*	0/1.4*	0/0.15*

3.1.1. Technology

3D direct laser writing is a rapid prototyping technique based on multiphoton absorption, enabling the fabrication of arbitrary 3D nanostructures in suitable photoresists such as IP-G 780 or SU-8 (Figure 3.1).[92] The photoresist is perfectly transparent to the laser light since the one-photon energy lies below the absorption edge of the material for the chosen wavelength. However, by tightly focussing the light of the ultrashort-pulsed laser, the intensity is high enough to expose the photoresist in the focal volume by multiphoton absorption. Within this small volumetric pixel ('voxel'), the absorption causes a chemical and/or physical modification of the photosensitive material. Using a developer bath, the unexposed regions are removed after writing.[95]

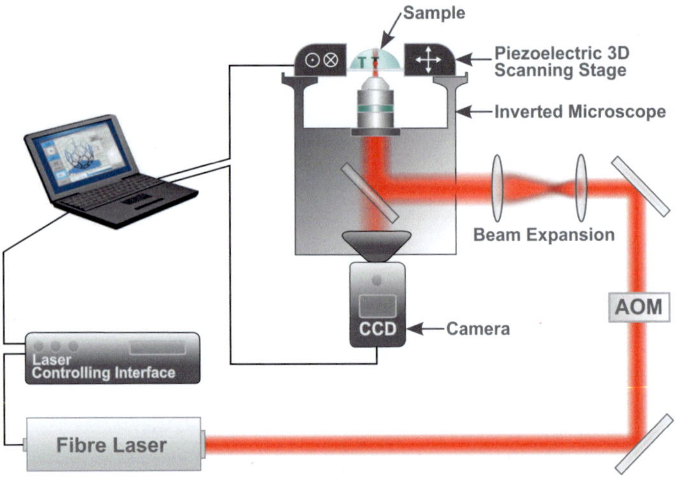

Figure 3.1.: Schematic of the 3D direct laser setup (Nanoscribe GmbH) used for this study. The beam of the ultra-short pulsed fibre laser is focussed into the photoresist by a high numerical aperture objective. The control of the laserpower with an acousto-optic modulator (AOM) allows scaling of the voxel. By moving the piezoelectric scanning stage, the sample position can be shifted relatively to the fixed focal position, enabling the writing of arbitrary paths into the material.

3.1.2. Fabrication of Gecko-Inspired Arrays

3D direct laser writing allows for the rapid fabrication of arbitrary micro- and nanostructures. However, several challenges occurred, so that more than 10 attempts were necessary to cope with these issues (Figure 3.2). Frequently structures collapsed or peeled off from the substrate. Furthermore, the correct exposure doses had to be found in order to achieve the structure dimensions accurately. Additionally, the enormous amount of filigree structures (> 15,000) that had to be individually written by the laser increased the processing time beyond 50 hours. In the end, appropriate process parameters were found to achieve the wished structure dimensions precisely. In the following, the discovered challenges and the final solution are reported.

Challenges in Fabrication. During the attempts, three different types of negative-tone photoresists were tested:

- SU-8 50,
- IP-L 780, and
- IP-G 780.

All these photoresists possess elastic moduli in the one-digit gigapascals range, making them comparable to the properties of bulk β-keratine, the material setae consists of.

First, the photoresist SU-8 50 was chosen which is very common and probably one of the most defined photoresists available in the market. During the initial attempts only the contour was exposed, meaning only the hull of the structures is written, which saved a lot of time. Theoretically, the unexposed liquid photoresist enclosed within the hull can be polymerized by flood exposure after development. Characteristically, SU-8 50 has a

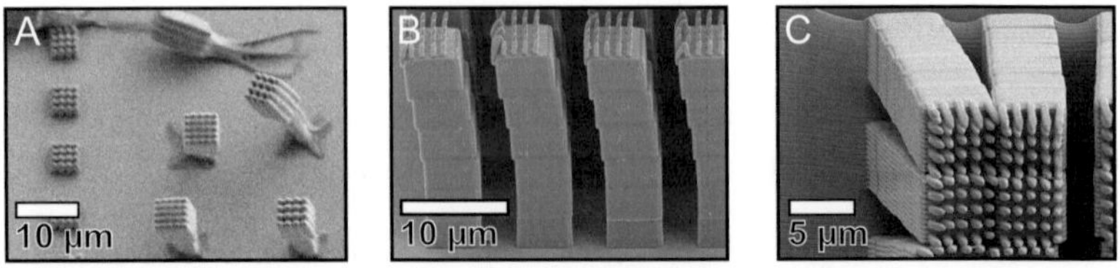

Figure 3.2.: Challenges in 3D direct laser writing. **A** The high shrinkage of SU-8 50 and the resulting inner tensions led to demolition of the structures. **B** Since the structures are written from top to bottom, floating of the resist results in aliasing. **C** During development, capillary forces caused structural collapse that even lifts the underlying socket.

comparatively low contrast, making the degree of polymerization extremely sensitive to the exposure dose. This would then easily allow the adjustment of the elastic modulus and therefore compliance of the structures by changing the degree of polymerization by applying another dose during flood exposure. However, the initial attempts revealed, that contour exposure results in insufficient stability leading to structural collapse due to capillary forces during development.

From this point on, the structures were fully exposed accompanied by an extension of the processing time to 50 hours and more. However, SU-8 shrinks enormously during polymerization leading to internal stresses. These stresses regularly forced the structures to collapse (Figure 3.2A). In consequence of these experiences, the IP-resists developed by Nanoscribe GmbH were chosen for the subsequent attempts. These photoresists are specifically designed for 3D direct laser writing and exhibit lower shrinkage and a low proximity effect. Yet, the liquid formulation IP-L 780 exhibits a low viscosity that led to floating of the written structures resulting in aliasing (Figure 3.2B). This is why the viscous gel formulation IP-G 780 soon came out on top.

However, the micro- and nanostructures written in IP-G 780 detached from the glass substrate during development due to capillary forces. In order to cope with this issue, freeze drying was applied to suppress capillary forces. To do so, the developer was exchanged by isopropyl alcohol which was substituted by cyclohexane afterwards. The liquid cyclohexane was frozen using an in-house cooling solution built by Felix Marshall.[96] After putting in a vacuum chamber, the frozen cyclohexane sublimates at pressures below about 5300 Pa.[96] However, owing to out-gassing the frozen cyclohexane was leaping up and down during sublimation. In this way, the exposed structures were damaged in all freeze drying tests.

Then, a thin layer of the adhesion agent OmniCoat (obtained from MicroChem Corp.) was coated between glass substrate and resist to prevent detachment from the substrate. However, the structures still detached. Next, a socket was added to the design in order to anchor the structures. Yet, the socket came off the substrate (Figure 3.2C). So, this idea was discarded and instead the substrate was coated with a 200 nm thin layer of SU-8. Before applying the IP-G 780 resist, the SU-8 film was soft baked and fully exposed. Using the SU-8 layer as adhesion agent worked very well and prevented the structures from detachment.

In order to accurately match the desired structure dimensions accurately, the correct dose had to be found for every unique voxel. The dose is mainly defined by the velocity of the laserspot and the laserpower. The more resist the laser light has to pass, the more laserpower is needed to keep the dose constant. Therefore, the structures were sliced into height sections (slices nanostrucutres: 360 nm, slices microstrucutres: 500 nm). By trial and error, the appropriate parameters were successfully found for every slice.

Successful Fabrication Process. For the fabrication of the presented gecko-inspired structures, a coverslip served as substrate for the photoresist. First, this coverslip was cleaned with isopropyl alcohol as well as acetone and blown-off with nitrogen afterwards. This preparation was followed by spin-coating a 200 nm thin layer of SU-8 (1. 500 rpm for 10 seconds with acceleration of 100 rpm/second. 2. 2600 rpm for

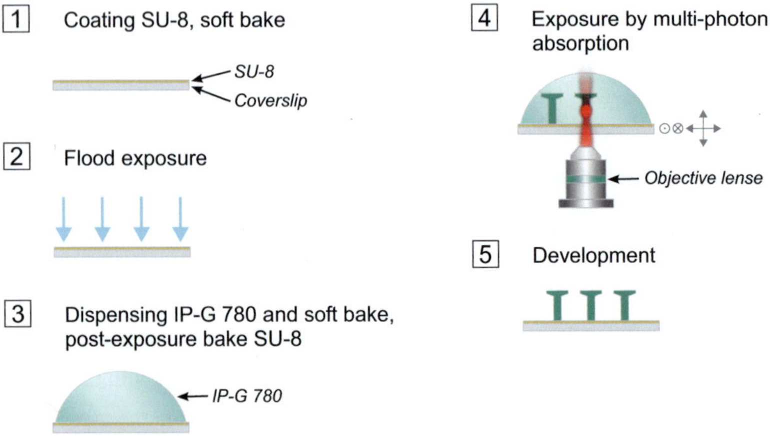

Figure 3.3.: Schematic of the five processing steps for the fabrication of gecko-inspired nano- and microstructures by 3D direct laser writing.

59 seconds with acceleration of 300 rpm/second). During the following processing, this thin layer of SU-8 on top of the coverslip ensured reliable bonding of the photoresist to the substrate. After soft baking (100 °C, 2 minutes), the SU-8 layer was exposed completely (6 minutes with a 36 W UV-source). Afterwards the chosen photoresist IP-G 780 was dispensed on top by using a pipette. Baking the compound for 90 minutes at 100 °C ensured the post-exposure bake of the SU-8 layer and soft-bake of the IP-G 780 resist. Subsequently, the sample was inserted into the laser lithography system. After writing completion, the structures were developed for 30 minutes in a PGMEA developer bath, followed by a 5 minutes cleaning in isopropyl alcohol and drying by nitrogen. The performed processing steps are summarized in Figure 3.3.

Figure 3.4 and Table 3.1 summarize the structures and dimensions fabricated and investigated in this study. Depending on the actual set of parameters, the resulting quadrangular arrays have an edge length between 68 µm and 82 µm. The structure's dimensions were verified by scanning electron microscopy at tilt angles of 0° and 70°. The width of the square

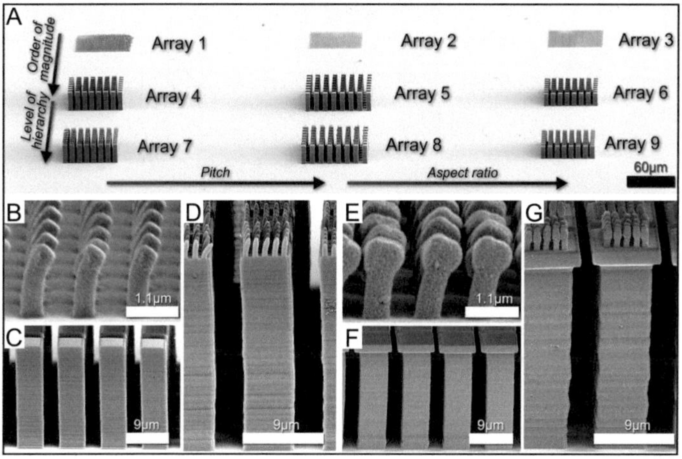

Figure 3.4.: A By using 3D direct laser writing, several arrays of gecko-inspired structures were fabricated. The arrays in the left column are optimized with regard to their aspect ratio and density according to the criteria of robust gecko adhesion (Chapter2). The density is reduced in the middle column and the aspect ratio is reduced in the right column. In addition, the smaller and larger pillars were combined to get a two-fold hierarchy (array 7, 8, and 9) and changed the tip shapes. **B** SEM image of array 1, which contains small single level structures with a width of only 500 nm. **C** The SEM image shows array 4, which contains pillars of 5 µm width. The dimensions are 10 times larger than they are in array 1. **D** Array 7 consists of the array 1 on top of array 4. **E, F** and **G** show corresponding arrays with mushroom-shaped tips. All scale bars refer to the periodicity of the arrays (parameter p_1 in Table 3.1).

pillars ranges from 500 nm to 5 µm. The periodicities of the pillars are between 1.8 and 2.8 times their width, whereupon the aspect ratios reach values up to 4.5.

Elastic modulus of the IP-G 780 Resist. The gecko's setae consist of beta-keratine with an elastic modulus of 1 –4 GPa.[45,56,57] It is only the hierarchical design of these setae that leads to a very low effective modulus near 100 kPa.[45] Building these setae out of a stiff material has many advantages, like the prevention of structural collapse despite high density and

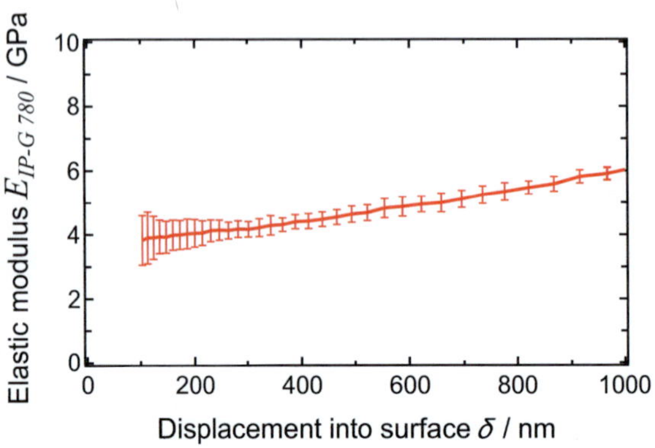

Figure 3.5.: Nanoindentation experiments were performed to characterize the elastic modulus of the used photo resist IP-G 780. The figure shows the calculated elastic modulus from CSM measurements. A common observation for soft and compliant thin films on harder and stiffer substrates, also seen here, is the rise of the elastic modulus with increasing indentation depth. A common assumption for this is that up to 10 % of the total film thickness, the measured values correspond to the real film properties. Hence, the elastic modulus of IP-G 780 is approximately 4 GPa, measured between 100 nm and 200 nm indentation depth.

aspect ratio of the fibers.[57] Hence, a stiff material was chosen to accomplish gecko-mimicking structures by 3D direct laser writing. The elastic modulus of the used photoresist IP-G 780 was determined by nanoindentation (Nano Indenter XP; Agilent Technologies, Santa Clara, CA, USA) performed by Sven Bundschuh (Institute for Applied Materials (IAM), KIT). The indentation experiments were conducted using the dynamic contact module (DCM) with a Berkovich tip. The samples were indented with a constant strain rate of 0.05 s^{-1} to a depth of 1000 nm. In addition to the load and displacement data, the instrument provides information on the contact stiffness continuously during the loading process via a superimposed displacement oscillation of 2 nm at 45 Hz (continous stiffness method, CSM).[97] By means of the measurement in Figure 3.5, the elastic modulus is determined to 4 GPa.

3.2. Adhesion Measurements

In the past, friction measurements were performed on hierarchical structures with a similar elastic modulus, however, the structures appear to be stochastic and their normal adhesion was not investigated.[98–100] A stochastic arrangement of gecko-inspired structures is disadvantageous for real world applications since synthetic adhesive tapes have to ensure a homogeneous adhesion all over the tape. In the following, a normal adhesion analysis and investigation of the impact of design variations on the adhesion of gecko-inspired structures with an elastic modulus close to the gecko's setae is presented. Due to the comparable small areas of the available arrays, the adhesion measurements had to be performed by atomic force microscopy (AFM).[101]

3.2.1. Procedure

Force-versus-distance measurements allow for the investigation of surface interactions and properties, i.e. adhesion force, separation energy, and compliance (Figure 3.6A). The tip apex radius of conventional AFM cantilevers, however, is in the range of nanometers, and much smaller than the smallest lateral dimension of the fabricated samples (Figure 3.6B). Therefore, a spherical silica particle of about 20 µm in diameter was mounted on a tipless AFM cantilever (Figure 3.6C).[102] This set-up also eliminates the need for a complex alignment procedure, required for a flat probe.[103]

So-called force maps were recorded for investigating the adhesion of the fabricated structures (Figure 3.7). Each force map covered a minimum area of 25 µm × 25 µm and contained at least 1024 (= 32 × 32) force-versus-distance curves. To analyze the adhesion, at least four force maps were measured for a preload of 100 nN, 500 nN, 1 µN, 2 µN, 3 µN, 4 µN, 5 µN, 6 µN and 6.7 µN. The averaged adhesion force was additionally referenced to the cross-sectional area of the spherical silica particle in order to identify

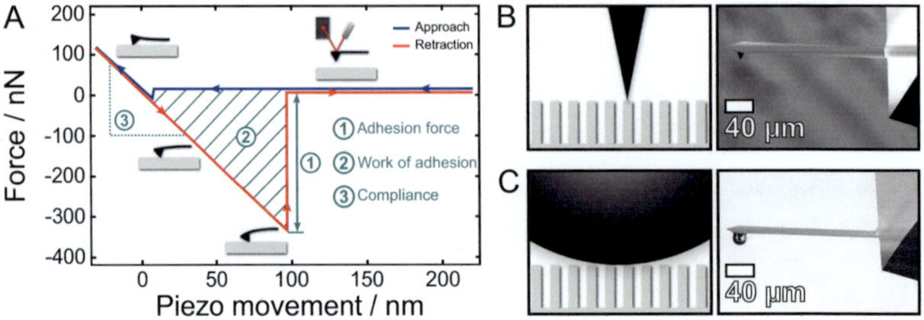

Figure 3.6.: A A force-versus-distance measurement comprises the approach and retraction of an AFM cantilever. If the cantilever is close enough to the sample surface, it feels the interaction forces. The resulting force-versus-distance plot reveals the adhesion force, the work of adhesion and the compliance. **B** Since the tip apex radius of conventional cantilevers is much smaller than the smallest lateral dimensions of the fabricated samples, they are not suitable for measuring the adhesion of these structures. **C** Instead of using sharp tips, the adhesion measurements were performed with a spherical silica particle attached to a tipless cantilever. Using a spherical particle eliminates complex alignment control.

the contact strength. To automate the analysis, custom software was coded to process all the relevant data of a force map (Figure 3.8).

The velocity during approach and retraction of the cantilever has to be chosen carefully, since visco-elastic effects are influencing the pull-off

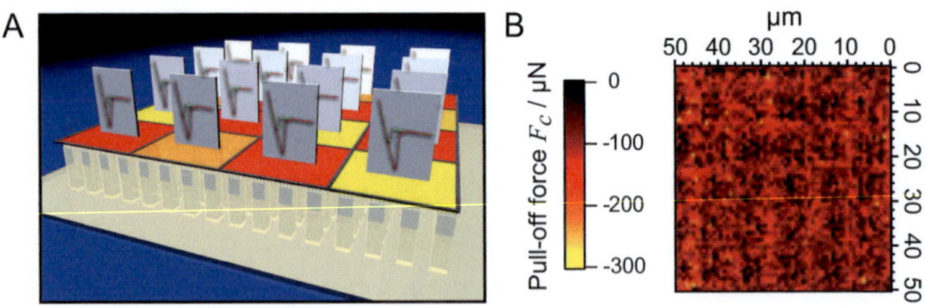

Figure 3.7.: A For measuring a force map, the sample surface is divided into several pixels. On every pixel a force-versus-distance curve is measured. After analyzing the measurements, the adhesion is shown as a color contrast in the force map. **B** As an example, a force map of 25 µm × 25 µm is shown. The force map is based on 64 × 64 force-versus-distance measurements.

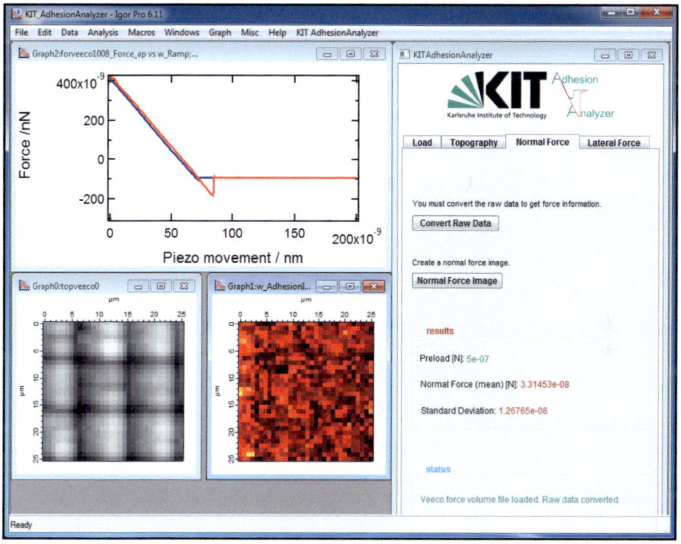

Figure 3.8.: Screenshot of the *Adhesion Analyzer* coded in the programming environment *Igor Pro* (Wavemetrics). The AFM data can import topography, force volume as well as lateral force images. After importing, the average force and standard deviation are displayed for the imported image.

force.[104] In this sense, the adhesion of selected arrays was measured as a function of the velocity. As expected, the adhesion has an upper boundary value that is reached for velocities faster than $v = 12\,\mu\text{m}\,\text{s}^{-1}$ (Figure 3.9).

Analyzing the contact of a flat IP-G layer with the spherical silica probe allows for the estimation of the contact area during the measurements. By the inspection of the Tabor coefficient for the given contact, it can be decided whether the JKR model or the DMT model has to be applied.[73,105,106] The JKR model leads to more realistic description of the contact between large and soft solids, whereas the DMT model is more suitable for small, hard solids.[107] Supposing the reduced radius of curvature $C_f = 10\,\mu\text{m}$, the elastic modulus $E_f = 70\,\text{GPa}$, the Poisson's ratio $\nu_f = 0.17$ for the silica probe and $C_s = \infty$, $E_s = 4\,\text{GPa}$ and $\nu_s = 0.22$ for the IP-G layer

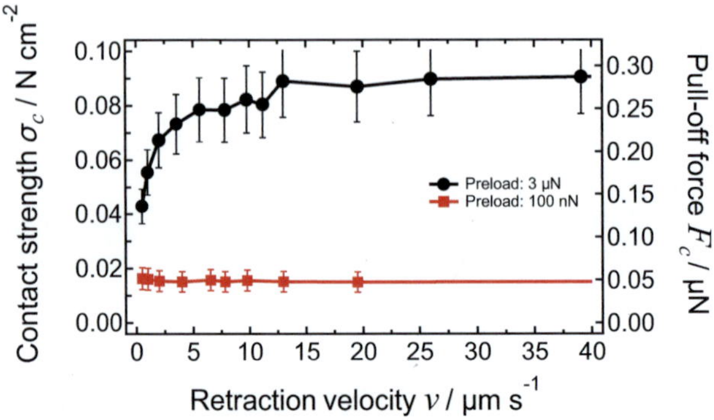

Figure 3.9.: The adhesion has been measured as a function of the retraction velocity. For velocities faster than $12\,\mu\mathrm{m\,s^{-1}}$, the adhesion reaches its upper boundary value.

and assuming a surface energy of $\gamma = \gamma_f = \gamma_s = 45 \times 10^{-3}\mathrm{Nm}^{-1}$, the Tabor coefficient is

$$\mu = \frac{h}{Z_0} = \left(\frac{\gamma^2 C^*}{Z_0^3 E^{*2}}\right)^{\frac{1}{3}} \approx 47.5, \tag{3.1}$$

where h is the neck height around the contact zone and Z_0 is the equilibrium separation of the atoms with a typical value of 3 Å.[105] C^* represents the radius of curvature of the silica probe and E^* the reduced elastic modulus as defined by[73]

$$\frac{1}{C^*} = \frac{1}{C_f} + \frac{1}{C_s} \quad \text{and} \quad \frac{1}{E^*} = \frac{1-\nu_f^2}{E_f} + \frac{1-\nu_s^2}{E_s}. \tag{3.2}$$

Since $\mu > 1$, the JKR model is valid for the given contact.[107] The real contact radius between a sphere and a flat surface is given by[73]

$$r^3 = \frac{3C^*}{4E^*}\left(F + 3\pi W C^* + \sqrt{6\pi W F C^* + (3\pi W C^*)^2}\right), \tag{3.3}$$

where r is the real contact radius, W the work of adhesion and F is the external force. With $W = \gamma_f + \gamma_s = 2\gamma$ this leads to a contact area with a diameter between 700 nm and 820 nm for the given preload range. The contact area at pull off is[108]

$$r_c = 0.63 \left(\frac{6\pi W C^{*2}}{E^*} \right)^{\frac{1}{3}} \approx 350\,\text{nm}. \tag{3.4}$$

The largest real contact area during a force-versus-distance measurement is, therefore, at the maximum compressive load and is expected to be $< 1\,\%$ than the cross sectional area of the spherical probe.

A significantly reduced real contact area compared to the apparent contact area is characteristic for rough contacts. The simplest rough surface could be imagined as a surface uniformly covered by asperities which have all the same radius of curvature C and the same height (Figure 3.10A).[109] The center of all asperities in Figure 3.10A is at the position $z = 0$, however, real surfaces usually have a random roughness. The model of Greenwood and Williamson[110] enhances the previously described model with randomly distributed heights of the asperities (Figure 3.10B). The center of each asperity n is displaced by Δd_n from the mean plane of center, whereby the distribution of Δd_n is Gaussian.

While performing the adhesion measurements, a spherical probe ($\varnothing = 20\,\mu\text{m}$) was pressed into the sample with a given preload. The relative position d of the spherical probe while performing the measurements is exemplarily shown in Figure 3.10C (measurement of array 3 with mushroom-shaped tips; preload: 6 µN). The corresponding deviations of the center mean position Δd_n is Gaussian distributed. Assigning a summit with radius of curvature $C = 10\,\mu\text{m}$ to each height position d_n leads to a rough surface topography with the root-mean-squared roughness of $R_q = 8.7\,\mu\text{m}$. The distribution of Δz_m, the deviation of the height from its mean height, is shifted in a way that is characteristic of rough surfaces with high peaks. Thus, the performed adhesion measurements approximate the contact of the sample with a micro-rough surface.

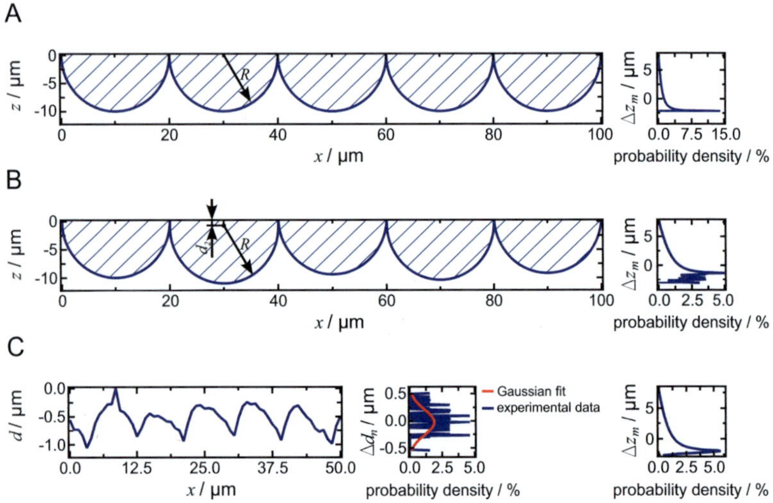

Figure 3.10.: **A** The simplest roughness could be imagined as uniformly distributed asperities which have all the same radius of curvature C and the same height. Since the center of the asperities are all at the same height instead of Gaussian distributed, such a surface does not describe a roughness in the proper meaning of the word, but rather a profile. The height probability density referring to the mean height of the profile is shown on the right. **B** According to Greenwood and Williamson,[110] a roughness can be modeled by asperities whose summits have all the same radius of curvature C, but with random height. The corresponding height probability density becomes Gaussian distributed. **C** In the adhesion measurements that were performed, the spherical probe touches the sample at different heights. The relative heights are exemplarily plotted as the relative height d vs. the lateral coordinate x. The distribution of the relative height is Gaussian, as required by the model of Greenwood and Williamson. Assigning a summit with radius of curvature $C = 10\,\mu\text{m}$ to each height position leads to a rough surface topography with a skewed height probability density that is typically for rough surfaces with high peaks. Hence, the adhesion measurement with a spherical probe imitates the contact with a rough surface.

3.2.2. Results

In order to investigate how geometry affects the adhesion of gecko-like adhesives, contact strength versus preload plots were recorded. For the plots shown in Figure 3.11, at least four force maps were averaged. As a reference, all plots contain the adhesion vs. preload curve (crosses) of a flat sample surface that has been fabricated by 3D direct laser writing, as well. As expected, the measured adhesion of this flat reference sample is very low and exhibits basically no preload dependency. Compared to the cylindrical pillars, the pillars with mushroom-shaped tip show an improvement of the adhesive properties in all our measurements. This mushroom-shape does not only increase contact area, it also improves the stress distribution in the contact area during detachment as recently discussed by Carbone *et al.*.[55] By the example of the two fold hierarchical structures, Figure 3.11A depicts the increased adhesion of the mushroom-shaped tips (solid circles) compared to the unstructured pillars (open circles). The improvement is seen for all preloads.

In Figure 3.11B the impact of density and aspect ratio is exemplified for the smaller pillars with mushroom-shaped tips (solid squares). A reduction of the density of about 20 % (solid stars) causes a negative effect comparable to a decrease in aspect ratio of 10 % (solid diamonds). This leads to an average loss in adhesion of 40 % for less density, and of even 50 % in case of a decreased aspect ratio. These slight modifications of the structures already demonstrate the strong influence of the design on the actual adhesion. However, one of the challenges in fabricating dry adhesives is the prevention of a structural collapse while increasing density and aspect ratio in order to improve the adhesion.[72,89]

In Figure 3.11C the effect of dimension and hierarchy is analyzed. Interestingly, the small single-level array (solid squares) exhibits the highest adhesion performance for all preloads compared to the larger single-level structures (solid triangles) and the hierarchical design (solid

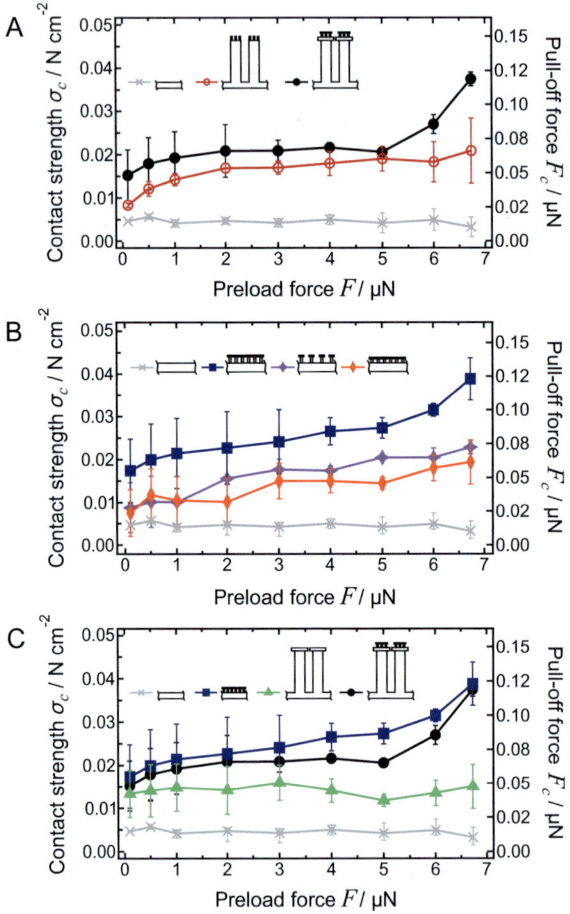

Figure 3.11.: The strong dependence of the adhesion on the actual design of the gecko-mimicking structures is revealed by the shown contact strength versus preload plots. Open symbols correspond to unstructured pillars while solid symbols refer to mushroom-shaped tips. **A** The positive impact of the tip shape on the adhesion values is shown for the two-fold hierarchical structures with (solid circles) and without (open circles) mushroom-shaped tips. **B** The influence of density and aspect ratio is analyzed in this graph. The decrease of density (−20 %; solid stars) and

aspect ratio (−10 %; solid diamonds) causes a significant reduction in adhesion by 40 % for a lower density and 50 % for a reduced aspect ratio. **C** In this plot, the effect of the hierarchical design is investigated. The small single-level structure (solid squares) exhibits consistently the highest adhesion, closely followed by the hierarchical design (solid circles). The difference in the adhesion of those two arrays is probably induced by the larger area fraction of the small single-level structure. However, at higher preloads the hierarchical structures become more and more compliant whereby their lower area fraction is compensated.

circles). However, in particular for higher preloads, the adhesion of the hierarchical array increases greatly and finally reaches the same adhesion as the small single-level array. Unfortunately, the properties at higher preloads could not be investigated due to the relatively low spring constant of the cantilever ($6.9\,\mathrm{N\,m^{-1}}$). The lower adhesion performance of the hierarchical array up to a certain preload, presumably results from its smaller area fraction. For small preloads, the contact area is dominated by this area fraction. In this case less area fraction, therefore, leads to less adhesion. However, at higher preloads the larger pillars of the hierarchical structures can bend and so the compliance increases which results in increasing contact area and therefore higher adhesion values.

The JKR-area during testing is predicted to be between 700 nm and 820 nm in diameter and hence the adhesion of the larger single-level pillars is curious at first sight. The area on top of the considered pillars is larger than the calculated JKR-area during testing. Hence, the measured adhesion of the larger single-level pillars should be comparable to the adhesion of the flat reference sample. The reason for the increased adhesion is revealed by the analysis of the adhesion maps shown in Figure 3.12C: the adhesion on top of the large single-level structures is very low and comparable to the adhesion measured on the flat reference sample, but is high at the structure edges. Since the used force measurement technique is not capable of distinguishing between different interactions, the high adhesion at the structure edges is mainly attributed to mechanical interlocking. The diameter of the spherical probe was small enough to sink into the gaps between the pillars. Consequently, a negative frictional force is detected during retraction due to clamping effects which is not to be confused with the perpendicular adhesion force originating from van-der-Waals forces. Figure 3.12 additionally shows the modification of the adhesion by adding the small single-level structures on top of the large single-level structures. This configuration leads to uniformly distributed adhesion all-over the array.

Typically, the adhesion of geckos is anisotropic, meaning the pull-off force necessary to detach a seta strongly depends on the direction of the preload. Preloading in both the perpendicular and parallel direction results in an over 10 times higher pull-off force of a single seta compared to preloading a seta just in the perpendicular direction.[22] Accordingly, geckos tend to pull their feet inwards toward their body causing a parallel preload of the setae. The resulting micron-scale displacement is necessary to align the setae, maximize contact area with the substrate and activate adhesion.[46] Autumn *et al.*[111] further demonstrated significant adhesive friction when isolated setae arrays were dragged along their natural curvature. In contrast, when dragged against their natural curvature much less friction was exhibited and the adhesives pads were easily peeled from the surface.

The work of Autumn and coworkers inspired groups all over the world to mimic the directional adhesion in order to obtain strong adhesion and easy release in synthetic gecko-like adhesives.[112–114] Various designs have been fabricated in order to switch adhesion. Angled fibers with angled tips made of soft polyurethane exhibited significant adhesion when loaded in one direction, and self-releasing behaviour when loaded in the opposite shear direction. Furthermore, Murphy *et al.*[84] demonstrated that the adhesion can be controlled by varying the shear displacement before loading in normal direction. Shear-induced unidirectional adhesion has also been shown for stooped nanohairs and vertically aligned nanotubes with angled

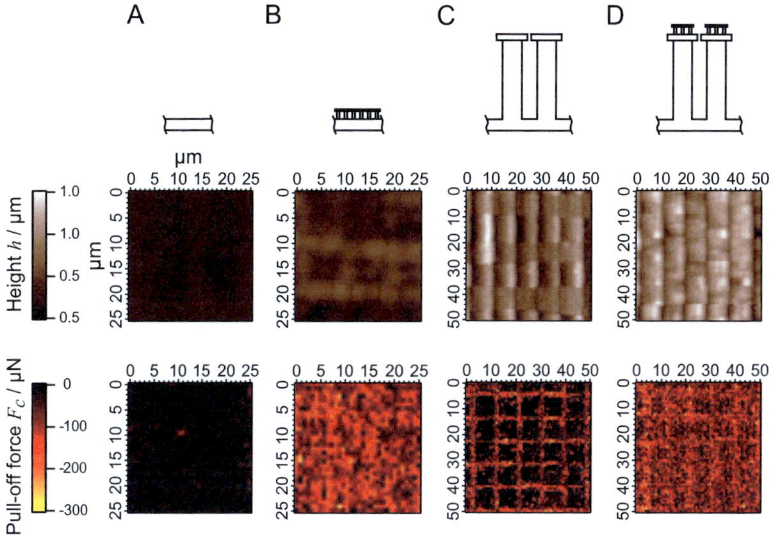

Figure 3.12.: Topography and adhesion measurements (preload: 6 µN) for the flat reference sample, the small and the larger single-level array and the hierarchical structure. The measurements were performed with an AFM using colloid tips with a diameter of 20 µm. **A** The topography image of the reference sample shows its flatness in the nanometer range (root-mean-squared roughness measured with the colloid tip $R_q = 35\,\text{nm}$) while the adhesion map highlights its poor adhesion.**B** Due to the large radius of the colloid tip the topography of the small single level structures is not properly imaged. The adhesion, however, is greatly enhanced compared to the flat reference sample. **C** The contour of the mushroom-shaped tips of the larger single-level array can be identified in the topography image. The corresponding force map, however, reveals that the tip frequently catches in the gap between the pillars. **D** This effect is prevented for the two-folded hierarchical structure. Since the probe cannot catch between the larger pillars, the hierarchical structure leads to uniformly distributed high adhesion.

tips.[83,115] Reddy *et al.*[116] demonstrated a fiber array made of a shape-memory polymer. Thermally induced, the authors showed switchable adhesion by tilting the fibers. Wedge-shaped designs and angled fibers benefit from their reduced stiffness and enhanced compliance.[24,32] However, the resulting pull-off force for angled fibers is less due to a rotational moment that is induced additionally.[32] Hence, in order to exceed vertical fibers in

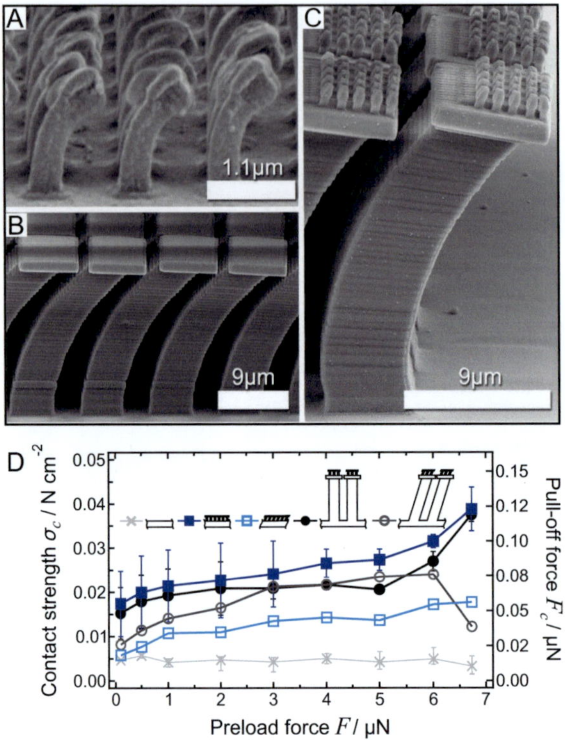

Figure 3.13.: **A** Array 1, **B** array 4 and **C** array 7 were fabricated additionally with tilted posts. **D** Compared to the straight arrays (solid symbols), an adhesion improvement could not be observed for the arrays with tilted pillars (open symbols).

terms of adhesion, angled fibers have to compensate the decreased pull-off force by an increased contact area. However, Reddy *et al.*[116] and Aksak *et al.*[32] measured less adhesion for angled fibers compared to vertical fibers which were preloaded perpendicularly only.

By using 3D direct laser writing, fabricated tilted posts with mushroom-shaped tips were fabricated in order to explore the various options given by this fabrication technology (Figure 3.13A-C). Subsequently, the tilted posts were examined with an AFM and the previously described method for measuring the adhesive forces. Since the testing protocol excludes lateral

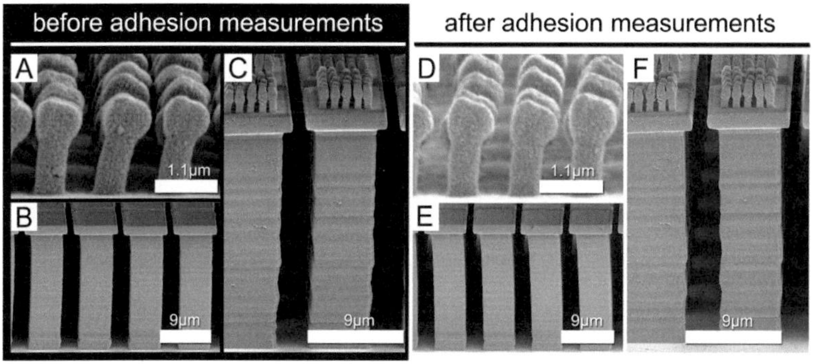

Figure 3.14.: The structures fabricated by 3D direct laser writing are very durable and showed no observable damage after thousands of adhesion measurements. **A** SEM image of array 1, which contains small single level structures with a width of 500 nm and mushroom-shaped tips. **B** The SEM image shows array 4, which contains mushroom-shaped pillars of 5 μm width. **C** Array 7 consists of array 1 placed on top of array 4. The SEM images shown in **A**, **B**, and **C** were taken before performing the adhesion measurements. **D**, **E** and **F** show the same arrays after the adhesion measurements.

movement of the cantilever, the structures were preloaded in perpendicular direction only. The adhesion vs. preload curve displayed in Figure 3.13D compares the influence of tilting on the adhesion for small single-level structures and two-fold hierarchical structures as well. In agreement with the work of Reddy *et al.*[116] and Aksak *et al.*,[32] the adhesion of tilted structures is less when being preloaded in perpendicular direction only. Adjusting the testing protocol in order to allow parallel preloads and measuring friction is future work.

Finally, the robustness of the structures written by 3D direct laser writing has to be emphasized. They survived thousands of force vs. distance curves taken with the AFM. Comparing SEM images of the structures before and after the adhesion measurements, no damaged or worn pillars could be observed (Figure 3.14).

3.3. Conclusion and Outlook

In conclusion, a systematic design study and adhesion analysis of gecko-mimicking structures is presented. These structures are close to the inspiring example of the gecko in dimensions and elastic modulus. With 3D direct laser writing it is straightforward to design, fabricate, and test arbitrary nano- and micrometer scale structures with or without hierarchy in a very flexible way. The interpretation of the presented results was supported by adhesion maps obtained with colloid AFM tips. The positive impact of mushroom-shaped tips could be demonstrated for stiff materials with lateral dimensions in the nanometer range. In addition, it was shown that the hierarchical structure of dry adhesives positively affects adhesion for appropriate preloads. This result supports the long-standing hypothesis that adhesion to natural rough surfaces requires a hierarchical design to ensure intimate contact and therefore a high overall amount of van-der-Waals forces.

In addition, 3D direct laser writing offers the possibility to fabricate three-dimensional templates, so that arbitrary but demoldable soft struc-

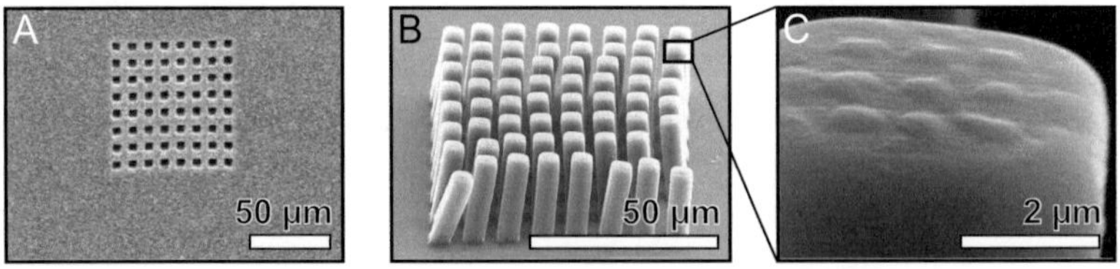

Figure 3.15.: SEM images of the fabricated rubber template and the molded polyurethane structures. **A** To replicate the structures that were created by 3D direct laserwriting, a template was fabricated by pouring a liquid silicone rubber (HS-II, Dow Corning). **B** Using this template, the master structure was replicated by vacuum-molding of polyurethane (TC-892, BJB Enterprises). After pouring and curing the polyurethane, the microfibers were carefully peeled off. **C** As shown in the magnified image, the nanostructures are hardly visible and were not molded successfully. Possibly, using other materials for this process could help to address this issue.

tures can be easily manufactured by casting (Figure 3.15). Therefore, the reported process will prospectively loom large in the fabrication of functional surfaces.

*A shortened version of this chapter was published as the article "3D Direct Laser Writing of Nano- and Microstructured Hierarchical Gecko-Mimicking Surfaces". M. Röhrig, M. Thiel, M. Worgull and H. Hölscher: Small **8**, 3009 (2012).*

4. Soft Molding and Dipping of Self-Cleaning Gecko-Inspired Adhesives

Geckos hairy attachment system is contradictory at first. On the one hand gecko toes are exceptionally adhesive and, on the other, dirt seems not to impair their adhesive strength. This does not match our everyday experience with tapes or sticky notes that lose their adhesion strength after the slightest contamination (Figure 4.1). Geckos, however, have to maintain their adhesion in a 'dirty' habitat like jungles or deserts. The impressive ability of self-cleaning their toes enables geckos to keep their toes sufficiently adhesive by simply walking or climbing. Interestingly, only a few studies are reported that investigate the self-cleaning ability of geckos.[12,13] Contaminated gecko toes regained 35 % of their initial clean adhesion during a cleaning sequence of 8 manual preload - shear motion - pull cycles.[12] Allowing geckos to peel off their contaminated toes themselves, their digital hyperextension ability enables them to recover up to 80 % of initial adhesion strength after only four natural steps.[13]

So far, gecko-inspired adhesives were demonstrated that compare favorably to the gecko in attachment strength,[34,43,83,117,118] however, no synthetic gecko-adhesive matched its natural counterpart in the ability to regain adhesion after contamination. The most successful synthetic gecko-inspired adhesive that can be cleaned without requiring any auxiliaries recovers only 33 % of its initial attachment strength after 30 cleaning cycles.[81] Having an initial clean shear adhesion of 8 kPa, the adhesion of this synthetic adhesive is only 4 % of the clean shear adhesion of geckos.[113]

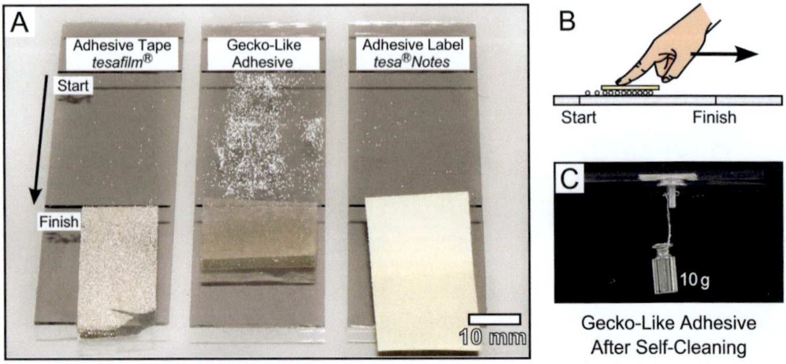

Figure 4.1.: A qualitative demonstration of the contact self-cleaning ability of the fabricated gecko-inspired adhesive. **A** An adhesive tape, the gecko-like adhesive and an adhesive label were contaminated with glass spheres. Thereupon the contaminated adhesives were manually dragged once over a glass slide as schematically shown in **B**. Both adhesive tape and adhesive label were only marginally cleaned of contaminants. In contrast, the gecko-inspired adhesive shown in the middle released nearly all contaminating glass spheres. For better visualization, the contrast of the released glass spheres is increased in this figure. **C** Dragging once was sufficient to clean the gecko-like adhesive enough to exhibit significant adhesion (adhesive pressure = 0.87 kPa, diameter of the specimen holder is 12 mm). Adhesive tape and label, however, did not even support their own weight since their adhesive zones were still fully covered by glass spheres.

To be practical, self-cleaning is required for most of the envisioned applications of gecko-inspired adhesives which are typically based on the reusability of such adhesives. Consequently, the development of self-cleaning gecko-inspired adhesives is one of the critical next steps to successfully produce biomimetic solutions such as: extremely strong, reusable adhesive tapes, non-irritating and reusable medical bandages,[119] robust climbing robots,[24,25] and industrial pick-and-place robotic manipulators.[26,27]

In this chapter, the mechanics of contact self-cleaning are investigated. Three sizes of mushroom-shaped microfibers were fabricated by combining soft molding and a dipping process.[32,84] By analyzing several size combinations of microfibers and contaminants three regimes of contact self-cleaning were revealed. Exploiting these findings, a synthetic gecko-inspired tape is presented combining high adhesive strength and self-cleaning ability that match geckos' properties very closely. As demonstrated in Figure 4.1 this exceeds by far commercial duct tapes or labels which only provide high adhesive strength at first but cannot be cleaned when contaminated. Inspired by the folds of skin on the gecko's foot (lamellae) contact self-cleaning is further improved by a hierarchical design. Finally, the essential rolling and sliding process of contact self-cleaning is explained in a theoretical model.

4.1. Fabrication of Mushroom-Shaped Microfibers by Soft Molding and Dipping

In order to examine the effect of size on contact self-cleaning, three types of gecko-inspired adhesive structures and a flat unstructured reference patch were fabricated by combining soft molding and dipping processes (Figure 4.2 and 4.3).[32,84]

Initially, an SU-8 master structure was fabricated with a two-stage UV-lithographic process. For this purpose, a thin layer of SU-8 (SU-8 2025, MicroChem) was spun on a glass wafer. After soft baking, the thin layer was uniformly exposed to UV-light followed by a post exposure bake. Afterwards, a second layer of SU-8 was spun on top of the exposed film. Subsequently, a chrome mask was used in order to UV-expose the master structure. After finishing this lithographic step, the unexposed regions were dissolved in a developer bath (SU-8 developer, MicroChem) revealing the micropatterned master structure.

In order to replicate this master structure, a soft template had to be created by pouring a liquid silicone rubber (HS-II, Dow Corning) over the SU-8 master structure. After curing for 24 h at room temperature, the template was peeled off the SU-8.

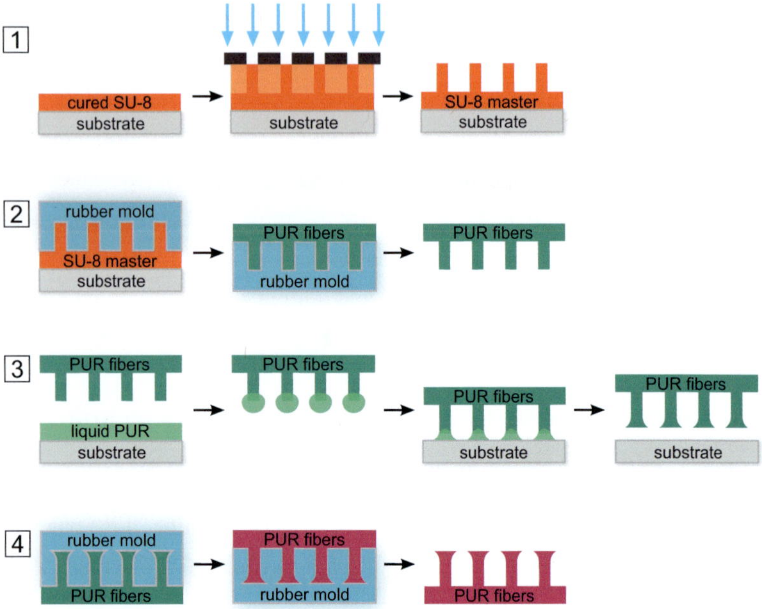

Figure 4.2.: Schematic overview of the fabrication of mushroom-shaped polyurethane fibers. **1** First, a master structure was created by UV lithography. **2** A template of the master structure was fabricated by pouring liquid silicone rubber. This template enabled the replication of the master structure by vacuum-molding. **3** Using a dipping process the replicated fibers received mushroom-shaped tips. **4** Creating a final rubber template allowed for the molding of mushroom-shaped microfibers.

This flexible template containing the inverse contour of the master structure served as a mold for the vacuum-molding of soft polyurethane (ST-1060, BJB Enterprises, elastic modulus: 2.9 MPa, work of adhesion to glass: $93\,\mathrm{mJ\,m^{-2}}$) After pouring and curing the polyurethane, the replicated microfibers were demolded by carefully peeling them off.

For the fabrication of widened microfiber endings, so-called mushroom-shaped tips, a dipping process was performed afterwards. For this purpose, a thin layer of liquid polyurethane was spun onto a polystyrene substrate.

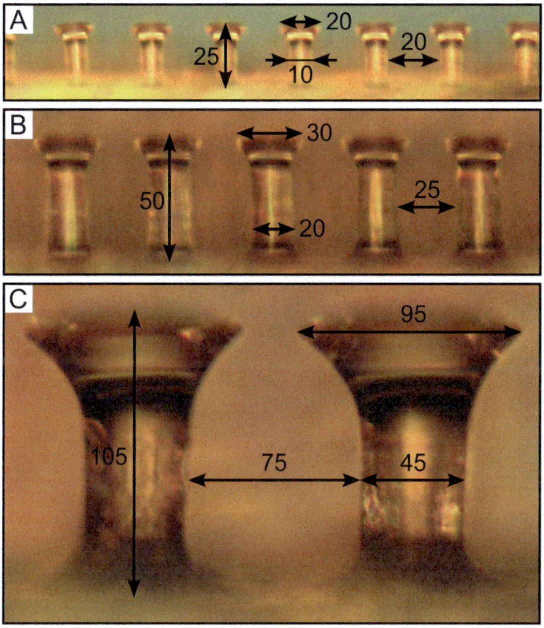

Figure 4.3.: Side-view optical microscope images of the three sizes of microfibers used in this study. All dimensions are in µm. **A** The small microfibers are 25 µm high and 20 µm wide at the tip. **B** The medium-sized pillars are 50 µm in height and 30 µm in diameter at the tip. **C** The large microfibers are 105 µm high and possess a tip diameter of 95 µm.

Briefly dipping the fabricated microfibers into the liquid polyurethane layer formed droplets at their tips. Pressing these endings onto a flat substrate resulted in mushroom-shaped microfibers after curing.

By fabricating a template with the previously described method these mushroom-shaped microfibers can be replicated by vacuum-molding. In this way, three sizes of polyurethane microfibers (tip diameter: 20 µm, 30 µm, and 95 µm) were fabricated (Figure 4.3).

4.2. Self-Cleaning Procedure

4.2.1. Contaminants

In order to investigate the contact self-cleaning of the fabricated microfibers, they have to be contaminated by well defined particles. Five distinct sizes of glass spheres ranging from 3 µm to 215 µm were used as contaminants (Table 4.1). In this way, the relationship of contaminant sizes to the dimension of the microfibers was investigated.

Table 4.1.: Measured diameters of the glass spheres used to contaminate the microfibers. Mean diameter $\varnothing_m$ and standard deviation $S.D.$ are given in µm. Both mean diameter and standard deviation were determined by analyzing the microscope images with customized software (Figure 4.4).

type	**supplier**	$\varnothing_m$	***S.D.***
GL–0191 1 – 15 µm	MO-SCI Specialty Products	2.8	1.7
SLGMS 45 – 53 µm	Cospheric	46.2	5.8
BBI–8541400	Sartorius	118.5	18.1
BBI–8541507	Sartorius	123.6	26.5
type 1922	Potters Industries	213.6	28.9

4.2.2. Testing Protocol of Self-Cleaning Measurements

The performed cleaning and adhesion measurements are divided into three phases (Figure 4.5):

1. adhesion measurement of the clean sample,
2. contamination of the sample, and
3. adhesion measurement followed by cleaning (alternating).

Below, the testing setup and the particular phases are discussed in detail.

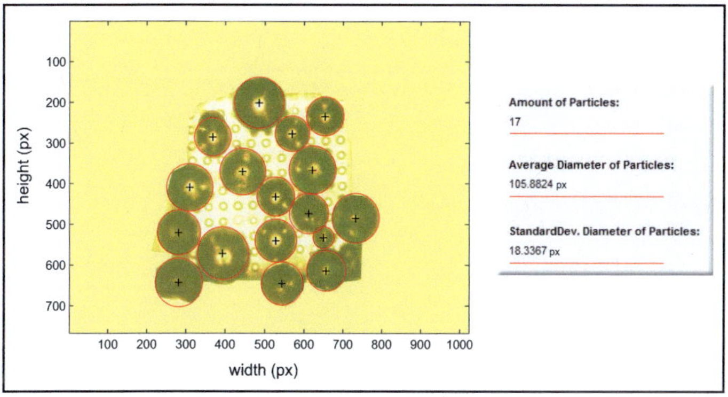

Figure 4.4.: Screenshot of the coded *Matlab* software that determines and analyses position and diameter of the contaminating particles. In this way, the contaminants were characterized in detail.

Setup. As shown in Figure 4.6, the testing protocol was carried out on a fully automated set-up that was built onto an inverted view optical microscope (Eclipse LE200, Nikon). Using in-house software, the custom 3-axis computer controlled motion control system (MFA-CC and VP-25XA, Newport) allowed the relative positioning of the microfibers and the glass slide serving as substrate (Microscope Slide, Pearl). The contact of the microfibers attached to the glass substrate was visualized by the inverted view optical microscope and a color digital video camera (DFW-X710, Sony). In order to generate reproducible results, the microfibers and the glass slide were properly aligned by manual adjustment of two rotational stages (GON40-U, Newport) and optical control. The normal forces applied to the adhesive sample were captured with a load cell and a signal amplifier (GSO-50 and TMO-2, Transducer Techniques) via a data acquisition board (NI PCI-6259, National Instruments). The commercially pre-cleaned glass slides that served as substrate were prepared for use by wiping with lint-free lens paper and blowing with compressed air. The glass microspheres used as contaminants were packed in dry air by the supplier and were used as is.

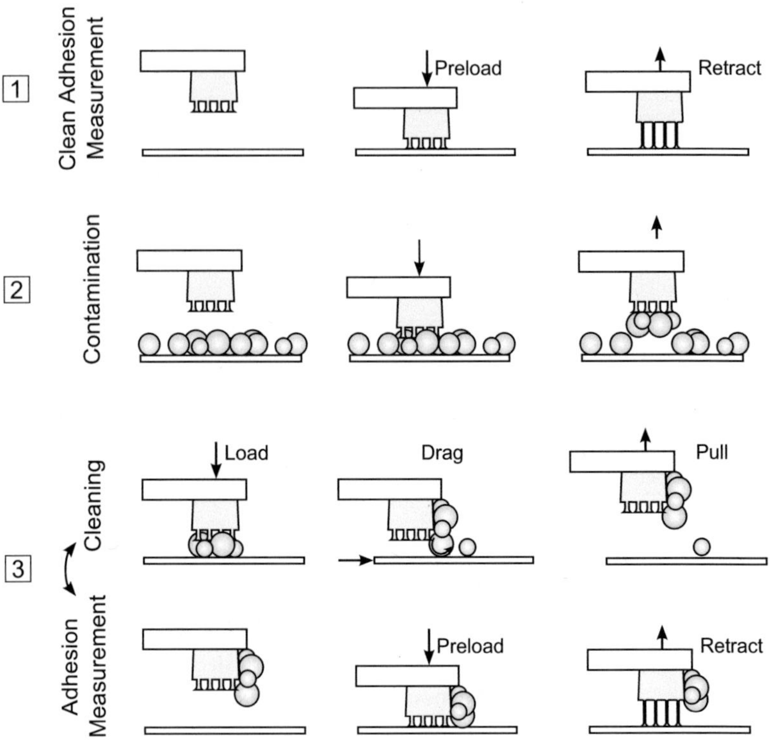

Figure 4.5.: Schematic of the performed cleaning and adhesion measurements.

Sample preparation. The samples had to be attached to the load cell via a cantilever. For this purpose, the samples were manually cut into squares (500 µm × 500 µm) by cutting with a scalpel along laser marked guidelines. Using double-sided tape, the cutted samples were mounted onto a clear acrylic peg which acted as a handling substrate. The peg itself was affixed to the clear acrylic cantilever with double-sided tape. Due to the manual cutting of the patches, edge structures were occasionally damaged. These errors were negated by the analysis protocol, which normalizes all self-cleaning results to the initial clean adhesion of each sample.

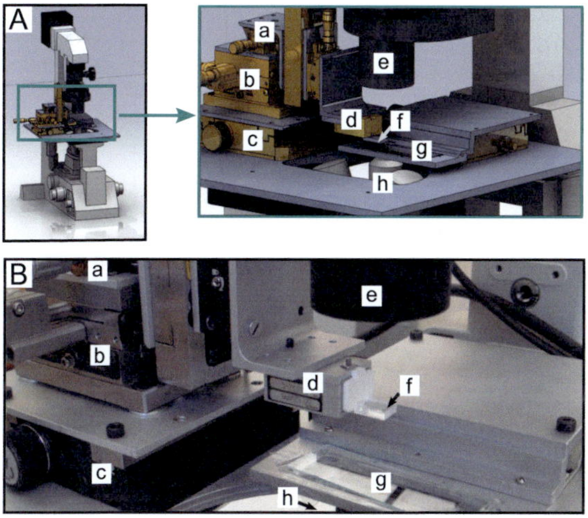

Figure 4.6.: **A** As this rendered model shows, the adhesion testing system was built on an inverted view microscope. **B** Photograph of the established testing setup. The labeled components are: **a** - goniometer, **b** - manual x and y axes stages, **c** - motorized y axis stage, **d** - load cell, **e** - light source, **f** - adhesive sample, **g** - glass contact substrate, **h** - microscope objective.

Clean adhesion measurement. According to Figure 4.5, the adhesion of the virgin samples was measured by vertically approaching the samples to the underlying glass slide until a predefined compressive load was achieved (velocity: $25\,\mu m\,s^{-1}$). Afterwards, the samples were retracted from the glass slide with the same velocity. The recorded forces were analyzed with in-house software (Matlab, MathWorks).

The resulting adhesion vs. preload plot comparing the data of all three different sizes of microfibers is shown in Figure 4.7. The large microfibers offer the highest adhesion (130 kPa, preload: 200 kPa, N = 10 samples). With decreasing microfibers, the measured adhesion drops and reaches 102 kPa (preload: 60 kPa, N = 5) for the medium microfibers and 85 kPa (preload: 40 kPa, N = 6) for the smallest microfibers.

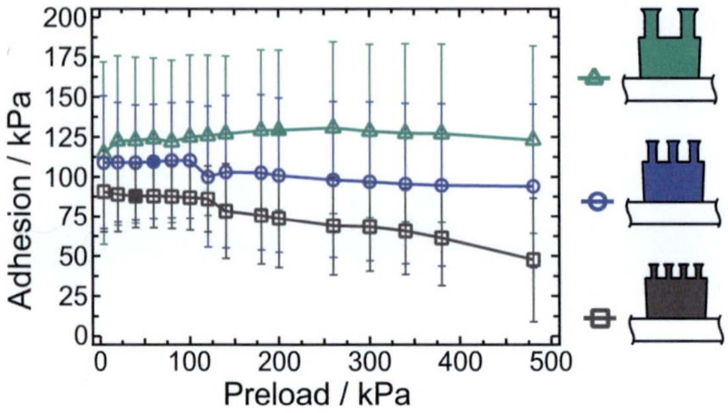

Figure 4.7.: Adhesion vs. preload measurements of the clean microfibers (large posts (N = 10 samples): triangles; medium posts (N = 4): circles; small posts (N = 5): quadrangles; total number of experiments: 855).

On the basis of these results the preload for the subsequent experiments was determined to preferably maximize the adhesion of each size of microfibers (filled symbols).

Contamination. After measuring the clean adhesion, each sample was contaminated by pressing it onto a monolayer of glass spheres (Figure 4.5). This was performed by pressing the sample onto the microspheres at a velocity of 25 $\mu m\,s^{-1}$ until a predefined compressive load was achieved, then the sample was retracted at the same velocity. According to Figure 4.7 the preload was determined to preferably maximize the adhesion (small sized structure: 10 mN (40 kPa); medium sized structure: 15 mN (60 kPa); large sized structure: 50 mN (200 kPa)).

To create a homogeneous monolayer of microspheres, one of the three following approaches was chosen: For microspheres larger than 150 µm in diameter, it was sufficient to manually pour them onto a glass slide, where they settled through gravity into a monolayer. Microspheres with diameters between 15 µm and 150 µm were poured onto a glass slide, then pressed with a glass cover slip to create a monolayer. For microspheres smaller than 15 µm in diameter, an aluminum surface was dusted with these spheres first.

Then, an electrostatic charge was built upon a glass slide by rubbing it with a piece of lint-free lens paper. Bringing the glass slide near the dusted surface, the spheres were attracted to the glass slide and formed a well-defined monolayer.

Adhesion measurements followed by cleaning. After contamination, the adhesion of each sample was measured immediately serving as the dirty case. To maintain a standardized cleaning procedure, the samples were cleaned by rubbing them against clean, dry glass slides. This was done through continuous load-drag-pull cycles with alternating dragging directions. In order to record how much adhesion was recovered, adhesion measurements were taken after each cleaning procedure and normalized to the initial clean adhesion (Figure 4.5). The direction of lateral displacement was alternated for each cleaning cycle to prevent any plastic deformation of the backing layer. The compressive load applied while dragging was empirically determined and identical for each post size (400 kPa). The glass slide serving as a substrate was cleaned as needed by wiping with a dry piece of lens paper and then with compressed air.

4.3. Self-Cleaning Analysis

Following the testing protocol described above, 24 different samples were used to perform more than 1000 experiments. In these studies all samples self-cleaned considerably (Figure 4.8). The rate and the degree of contact self-cleaning, however, significantly depended on the diameter of the contaminating particles and the tip diameter of the microfibers relative to one another. For classification, the non-dimensional parameter κ was introduced to represent the ratio between the contaminant diameter to tip diameter.

In this way, three regimes of contact self-cleaning (Table 4.2) were identified. In the *large contaminants regime* the diameter of the contaminating

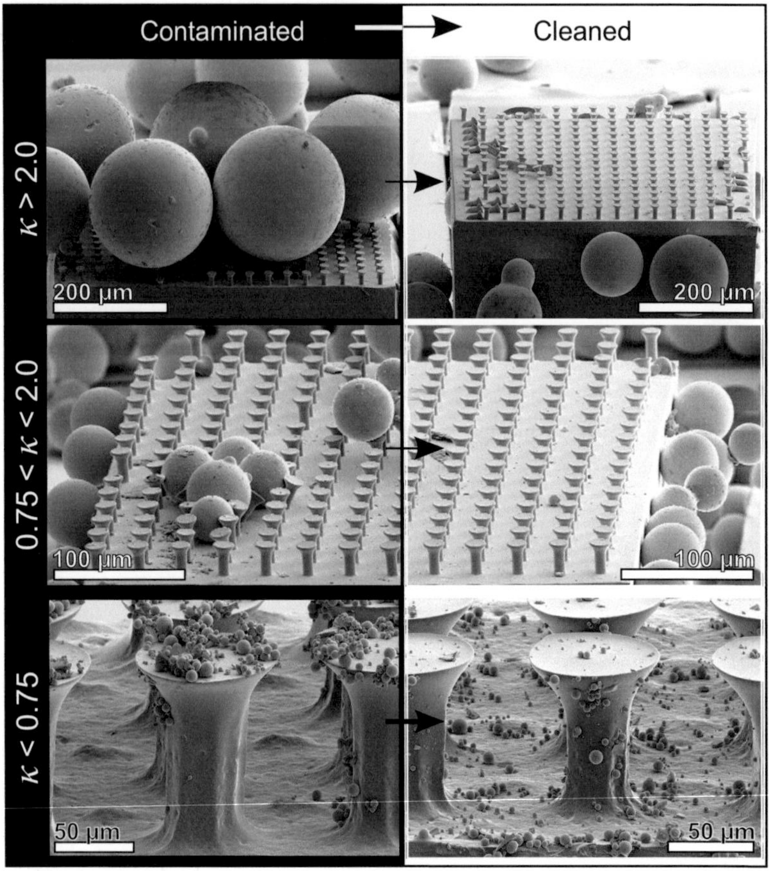

Figure 4.8.: The scanning electron microscopy images show the three self-cleaning regimes before and after dry self-cleaning.

particles is more than twice the tip diameters, i.e., $\kappa > 2.0$. In the *similar contaminants regime*, the value of κ is between 0.75 and 2. In case of contaminants that are smaller than 0.75 times the tip diameter the *small contaminants regime* is reached ($\kappa < 0.75$). In doing this classification different combinations of fiber and contaminant sizes are directly comparable. It is likewise valid to use another characterstic length, other than the tip diameter, for classification as can be seen from Figure 4.9.

D / B	Ø 2.8	Ø 46.2	Ø 118.5	Ø 123.6	Ø 213.6
25 / 20 / 20	0.14	2.31	5.93	6.18	10.68
50 / 30 / 25	0.09	1.54	3.95	4.12	7.12
105 / 95 / 75	0.03	0.49	1.25	1.30	2.25
	$\kappa < 0.75$		$0.75 \leq \kappa \leq 2$		$\kappa > 2$
	$D << B$		$D \approx B$		$D >> B$

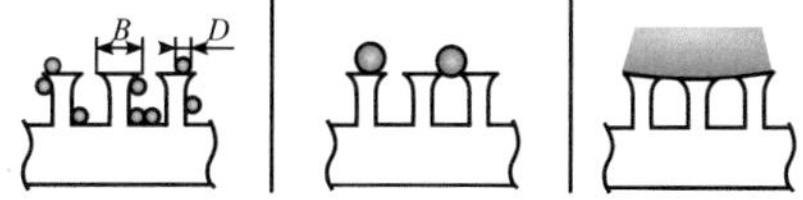

Table 4.2.: Classification of all self-cleaning regimes by listing all combinations of microfiber and contaminant dimensions. Three regimes were identified defined by their ratio κ of the microfiber tip diameter and the particle diameter: the small contaminants regime ($\kappa < 0.75$), the similar contaminants regime ($0.75 \leq \kappa \leq 2$), and the large contaminants regime ($\kappa > 2$).

4.3.1. Adhesion Recovery

In Figure 4.10 the adhesion recovery is plotted as a function of the performed cleaning cycles. The insertcd fits (solid lines) are based on the

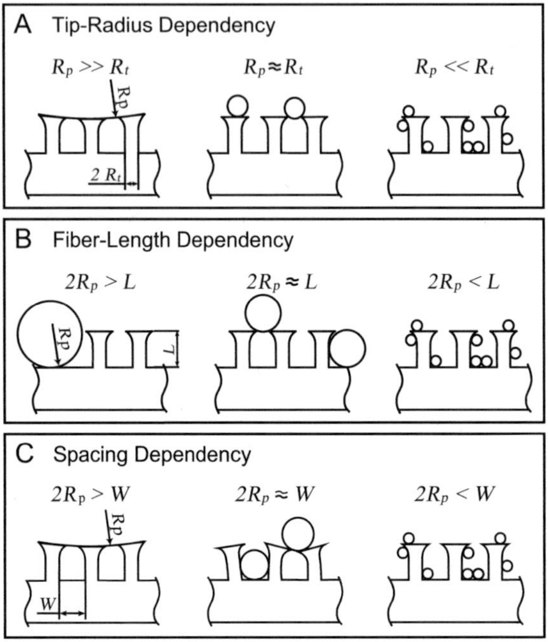

Figure 4.9.: Classification of self-cleaning regimes. The self-cleaning capability depends on the relative sizes of contaminating particles and the characteristic length of the microfibers. The characteristic length can be **A** the tip radius, **B** the length of the fiber, or **C** the spacing between fibers. In the case of tip-radius dependency three regimes were defined: the *big contaminants regime* for when the particle radius is greater than twice the tip diameter, the *similar contaminants regime* for contaminants that are about the same size as the tip diameter and the *small contaminants regime* for particles that are less than 0.75 times the tip diameter.

fitting function $P_{ad}(t) = F_{ad} - c\ exp^{-t/T}$, where P_{ad} represents the percentage of clean adhesion, F_{ad} the limit of adhesion recovery, c and T are the fitting coefficients and t is the number of cleaning cycles (Table 4.3). In doing so, the differences between the three contact self-cleaning regimes emerges. The capability of contact self-cleaning was maximized in the large contaminants regime (Figure 4.10A) whereas the rate and degree of self-cleaning are significantly reduced in the intermediate (Figure 4.10B)

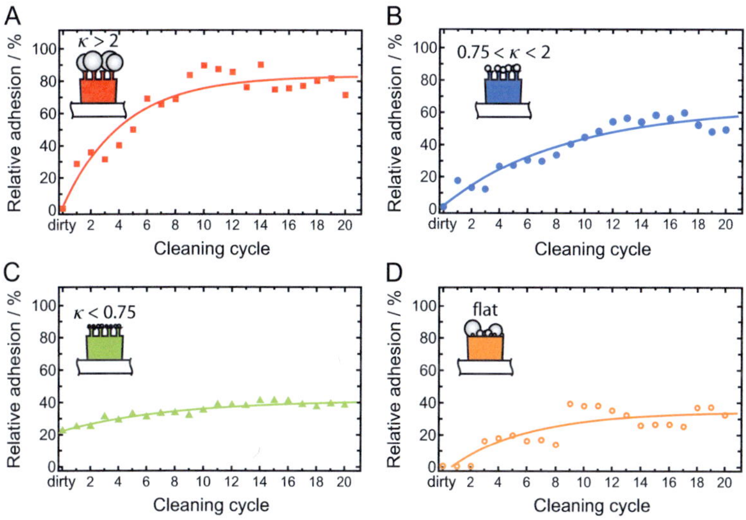

Figure 4.10.: Adhesion recovery of the three self-cleaning regimes and the flat reference patch. The regained adhesion of the samples are shown as a function of the cleaning cycles. To highlight adhesion recovery, the regained adhesion was normalized to the initial clean adhesion. The solid lines represent fits to the measured data. **A** In the large contaminants regime, the adhesion is nearly zero directly after contamination but recovers quickly to $F_{ad} = 83\,\%$. **B** In the intermediate regime rate and degree of adhesion recovery are lower ($F_{ad} = 63\,\%$). **C** The fibers contaminated by small particles retained some adhesion even when dirty, however, they did not recover much adhesion ($F_{ad} = 42\,\%$). **D** The flat control sample revealed the worst performance in both rate and degree of adhesion recovery ($F_{ad} = 35\,\%$).

and even more in the small contaminants regime (Figure 4.10C).

Self-cleaning in the large contaminants regime ($\kappa > 2.0$) saturated to 80 % of clean adhesion performance after only 9 cleaning cycles (with N = 5 investigated samples). The regime of similar contaminants ($0.75 \leq \kappa \leq 2$) regained 55 % of its initial clean adhesion within 12 cleaning cycles (N = 6). The lowest self-cleaning capability of the microfiber samples were observed in the small contaminants regime ($\kappa < 0.75$) that

Table 4.3.: Fit parameters for the three different contact self-cleaning regimes.

	c	T
$\kappa > 2$	81.2	4.3
$0 \leq \kappa \leq 2$	61.3	9.1
$\kappa < 0.75$	19.2	8.3
flat	38.2	5.6

saturated to 40 % within 11 cleaning cycles (N = 8). Interestingly, however, in contrast to the other two regimes, the adhesion directly after contamination was still about 20 % in this regime. This observation is considered in detail in the section below.

Flat unstructured adhesive patches of the same dimensions as the fiber-patterned samples (500 µm × 500 µm) served as control samples. In this study, these flat control samples were cleaned with the process parameters according to the ones used for the large posts. Only the largest particles used in the experiments were self-cleaned from the flat patch, however, even for these contaminants the measured self-cleaning performance was significantly behind all microfiber patches (Figure 4.10D). The adhesion recovery was only about 35 % after 20 cleaning cycles (N = 5).

4.3.2. Contamination-Resistance

The ability of the microfiber arrays to remain adhesive immediately after contamination, the so-called contamination-resistance, is investigated in the following (Figure 4.11). For this, the adhesion measurered directly after contamination but before cleaning cycles as well as the video data were analyzed. This analysis revealed that in the regimes of large and similarly-sized contaminants ($\kappa > 2.0$ and $0.75 \leq \kappa \leq 2$) almost no contamination resistance was exhibited since the contaminating particles blocked the contact between microfibers and substrate. However, for samples covered with small contaminants ($\kappa < 0.75$), adhesion remained at 22.4 % immediately after contamination (Figure 4.10C).

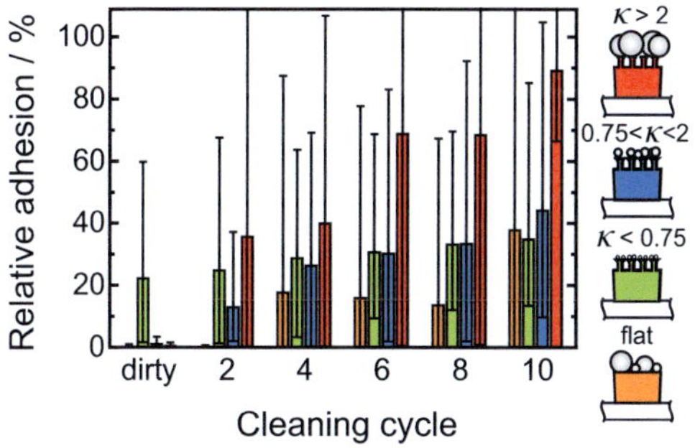

Figure 4.11.: Direct self-cleaning comparison within the first 10 cleaning cycles. Compared to the other regimes and the flat control the large contaminants regime demonstrates the most effective self-cleaning performance (red bars). Remarkably, only the microfibers in the small contaminants regime ($\kappa < 0.75$) maintain appreciable adhesion directly after contamination. The error bars of the measurements refer to the minimum and maximum value of the dataset.

Two factors are responsible for the observed contamination-resistance in the small particle regime. First, the contaminants are randomly distributed on the microfiber sample. Although they initially form a monolayer, the contaminating particles are stochastically distributed when applied to the microfibers and some fiber tips remain clean. Second, even when a fiber tip is contaminated the mushroom-shaped tip allows conformation around the particles as long as these particles are much smaller than the fiber.[53,55,120]

4.3.3. Self-Cleaning Modes

To investigate in which way the loading mode affects particle cleaning, the contaminants removed from the structures during both normal loading and shear loading steps were counted (Figure 4.12). Interestingly, 95 % of the removed contaminants were cleaned during shear loading and only the remainder during normal loading (S.D. = 13 %, N = 15). This indicates that shearing of the adhesives is very important for their cleaning.

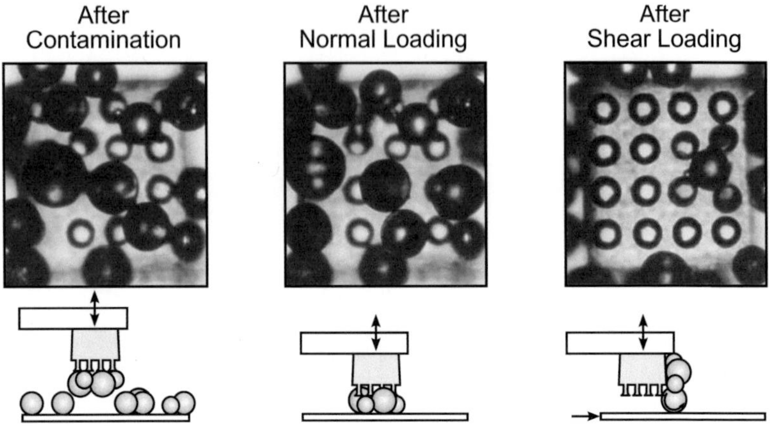

Figure 4.12.: Importance of shearing for contact self-cleaning. An example case of cleaning the large microfibers contaminated with particles that are 118.5 μm in mean diameter is shown. Normal loading did not remove any contaminants from the sample. However, the subsequent shear-loading step removed nearly all of the particles from the top of the patch.

Additionally, this study revealed two modes of contact self-cleaning. First, during contact self-cleaning the contaminants can be transferred to the clean substrate. However, this *deposition* seems to be less relevant than *embedding*. Here, particles are conveyed to the spacing between microfibers (Figure 4.13). Hence, regaining initial clean adhesion strength is not necessarily related to actually removing dirt from the adhesive. This result expands on previous studies, which only considered particles much greater in diameter than the fibers and only proposed self-cleaning by depositing[12,81] or rolling.[121] The principle task of geckos to rapidly recover adhesion is possible by just removing the contaminating particles from the tips of their setae. Where the dirt ends up on a short time-scale is not critical for the stickiness of geckos' toes. As shown in section 4.4, the self-cleaning performance of gecko-inspired microfibers can be further improved by considering the long-term deposition of contaminants by including lamellae-like structures.

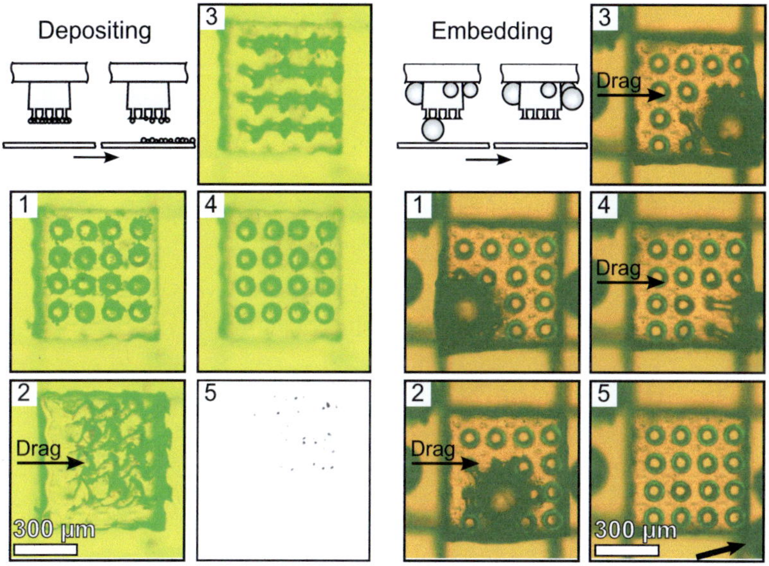

Figure 4.13.: Depositing and embedding are the most important modes of contact self-cleaning. **Depositing** is shown by means of an array of large microfibers (tip diameter: 5 µm) during one cleaning cycle. Due to the applied shear displacement the pillars collapse and the backing layer deforms. When released, the microfibers snap back. In this way, contaminants are deposited to the substrate as shown in the background-subtracted micrograph in frame 5. **Embedding** is shown using the example of a contaminating particle in the large contaminants regime ($\kappa > 2$). The applied shear force causes the contaminant to roll off of the microfibers. When reaching the end of the patch the contaminant is embedded to the backing as indicated by the arrow in frame 5.

4.3.4. Robustness

Shearing of the microfibers may cause physical damage and therefore decreased performance of the adhesives over time. In order to investigate self-inflicted damage due to contact self-cleaning, the adhesion of a clean patch of large microfibers was measured as a function of shearing steps. Even after 50 shearing steps, the initial adhesion changed only marginally, revealing the enormous robustness of the soft microfibers (Figure 4.14).

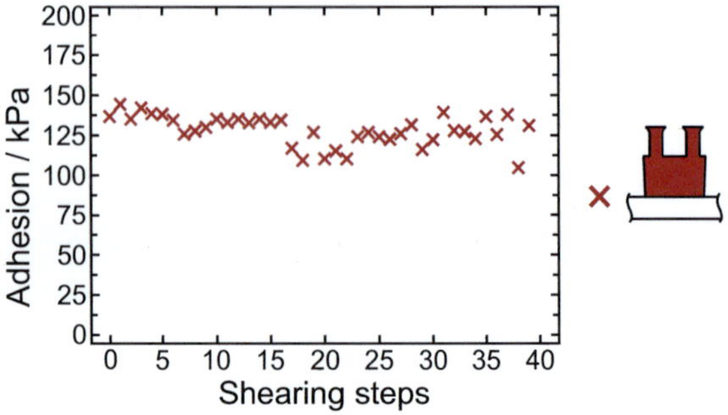

Figure 4.14.: To investigate to which degree wearing is an issue, 40 shearing steps were performed on a clean patch of large microfibers. After 40 cleaning cycles, the microfibers were still undamaged and their attachment strength only dropped marginally.

4.3.5. Synthetic vs. Gecko Self-Cleaning

The rate and degree of contact self-cleaning was maximized in the regime of large contaminants ($\kappa > 2$, Figure 4.10A). This leads to the conclusion that making microfibers much smaller than possible contaminants will result in synthetic gecko-like adhesives that are able to self-clean. As visual observations revealed, almost no contaminants were deposited directly onto the substrate. Instead, in the regime of large contaminants the particles were rolled along the microfibers until they reached the edge of the adhesive patch. Transferring these observations to the hairy attachment system of geckos indicates that the nanometer size of the spatulae does not only improve adhesion[52] but also ensures remaining in the large contaminants regime, while the stripy lamellae create cavities for the embedding of contaminants.

To make synthetic gecko-like adhesives applicable for real world applications, they have to be extremely adhesive in both their clean and cleaned states. The large microfibers (95 µm tip diameter) possess the highest clean adhesion among the three tested microfiber sizes. The large microfibers in the large contaminant regimes demonstrate up to 140 kPa of initial clean

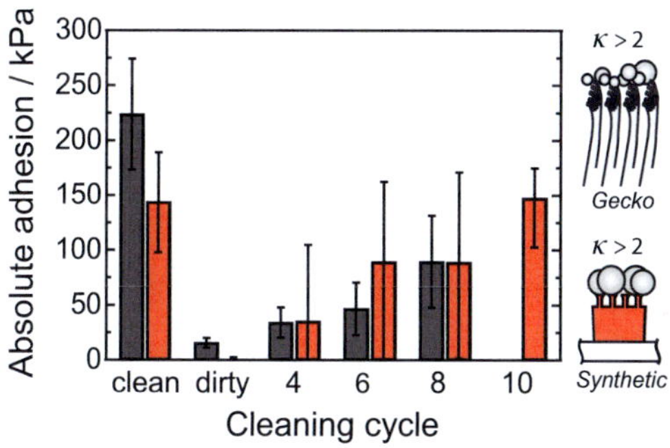

Figure 4.15.: Comparison of the absolute adhesion of geckos and synthetic microfibers. The measured adhesion recovery of large microfibers (95 μm tip diameter) in the large contaminants regime ($\kappa > 2$) eventually recovered to 100 % of the initial clean adhesion. In this case the synthetic gecko-like adhesive had similar quantitative performance to the gecko in terms of rate and degree of adhesion and contact self-cleaning (gray scale bars are data from literature[12]). The error bars that correspond to the synthetic microfibers refer to the minimum and maximum value of the dataset.

attachment force. Interestingly, they are not only comparable to geckos in absolute adhesion but also in their self-cleaning performance as shown in Figure 4.15, in which the observations of this work are compared with data of gecko setae published by Hansen *et al.*.[12]

For this comparison to be valid, similarities and differences in the testing protocols of this study of synthetic microfibers and gecko self-cleaning studies[12] have to be considered. The size of the microfiber samples used in this study was much smaller than the adhesive patch on a gecko toe. Consequently, the microfiber samples slightly benefit as adhesion seems to increase as patch size decreases.[53] In both studies, the cleaning protocols are similar as cleaning was conducted by shearing the adhesive along a clean glass surface. Furthermore, spherical particles in the large contaminants regime ($\kappa > 2$) were used in both works, although the particles were made of glass in this work, and of ceramic in the gecko study. Geometry and bulk material stiffness of the synthetic microfibers differ from gecko setae, however, their effective modulus is similar (gecko ≈ 100 kPa;[17] synthetic microfibers with 95 µm tip diameter ≈ 30 kPa, determined empirically). Finally, the adhesion of the microfibers acts perpendicular to the surface whereas the gecko's acts while shearing lateral to the substrate. Despite these differences, the principle comparison is valid and demonstrates for the first time a synthetic gecko-like adhesive that is appreciably as sticky as gecko toes and as efficient at recovering adhesion by contact self-cleaning.

4.4. Lamellae-Inspired Hierarchical Design

Inspired by the gecko's lamellae a hierarchical adhesive patch was created to exploit the observation that embedding is a primary means of self-cleaning and to qualitatively test this principle on a larger scale (Figure 4.16). By cutting out strips, this lamellae sample was patterned into rows of microfibers on raised ridges with intervening gaps (Figure 4.16A). In this way, an abstract analog of the gecko's lamellae was created. A sample of large microfibers (tip diameter: 95 µm) was the basis of this lamellae-inspired sample. By patterning, raised ridges that were four pillars wide (550 µm) and grooves half the width of the raised ridges were obtained, resulting in about 67 % of adhesive area coverage. After contaminating the lamellae-inspired samples by dipping into 110 µm diameter glass spheres, the sample was self-cleaned by manually shearing the lamellae patch along a glass slide (Figure 4.16C). This was enough to recover 17 % of its initial clean adhesion. In Figure 4.16E, however, it is shown that adhesion recovery is much more difficult in case of non-hierarchical samples as the contaminants have to roll along the whole area of the patch.

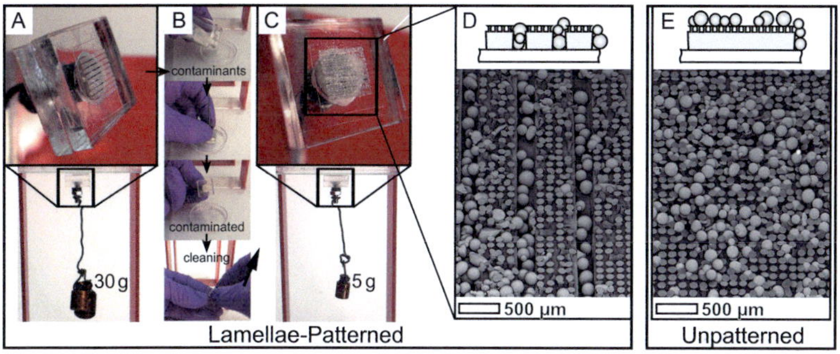

Figure 4.16.: Lamellae-inspired hierarchical design. **A** The stripy adhesive patch ($5.5 \times 5.5\,\mathrm{mm}^2$) of large microfibers was composed of 9 raised ridges, each 4 microfibers wide. These ridges were divided by gaps, half as wide as the ridges. Mounted onto a rigid acrylic backing the lamellae-patterned adhesive patch held 30 g of mass (about 10 kPa). **B** This series shows, from top to bottom: pouring of the 110 μm diameter glass spheres into a dish, pressing the adhesive into the spheres, the contaminated adhesive, and the cleaning process of rubbing the adhesive against a glass slide. **C** Although the adhesive patch appeared to be clogged with spheres even after cleaning, it regained enough adhesive strength to hold 5 g (1.7 kPa). **D** SEM imaging revealed that most of the contaminating glass spheres were trapped within the gaps. However, some were still contaminating the microfibers. Furthermore, some microfibers were destroyed due to the rubbing process. **E** On the other hand, the SEM images of the unpatterned patch of microfibers show that the contaminants remaining within the array impeded adhesion recovery.

As demonstrated in the qualitative proof of concept, adding grooves serving as collecting pans enhances the rate of self-cleaning for large-area patches as contaminants have not to cross the whole patch to be removed from the adhesive. However, two distinct limitations were observed. First, once the grooves are saturated with particles they can prevent the adhesive patch from properly contacting the substrate. Due to the non-optimal design of the lamellae-inspired patch, the rate and degree of contact self-cleaning will probably significantly lie behind the adhesion recovery of the small-scale adhesive patches. Additionally the stripy design leads to uneven shear pressure across the face of the adhesive surface during contact self-cleaning. This results in partly damaged structures due to stress concentration on the leading edges of the patch. Hence, a lamellae-inspired hierarchical design will enhance the rate of contact-self cleaning, however, the mentioned issues have to be addressed in the future. Possibly, this can be achieved by optimized groove designs and/or by mimicking the digital hyperextension of the gecko toe.[13]

4.4.1. Modeling

To gain more insight into the contact self-cleaning process, a model for a single spherical particle contaminating a fiber array is presented (Figure 4.17A). First, the critical shear force F_y necessary to achieve cleaning by sliding or rolling is calculated. For the contaminating particle to *slide* along the fiber array, the applied shear force must be higher or equal to the product of shear strength τ_f and contact area A_f of the fiber-contaminant interface[122]

$$F_y = \tau_f A_f, \tag{4.1}$$

where the contact area is assumed to be of circular shape ($A_f = \pi a_f^2$). Using the Johnson-Kendall-Roberts elastic contact mechanics theory,[73] the contact radius a_f is

$$a_f^3 = \frac{3R}{4E_{fc}^*}\left(F_z + 3\pi W_f R + \sqrt{6\pi W_f F_z R + \left(3\pi W_f R\right)^2}\right), \tag{4.2}$$

where R represents the radius of the spherical contaminant, E^* is the reduced elastic modulus of the fiber/contaminant system, F_z represents the applied normal load and W_f is the work of adhesion at the fiber-contaminant interface.

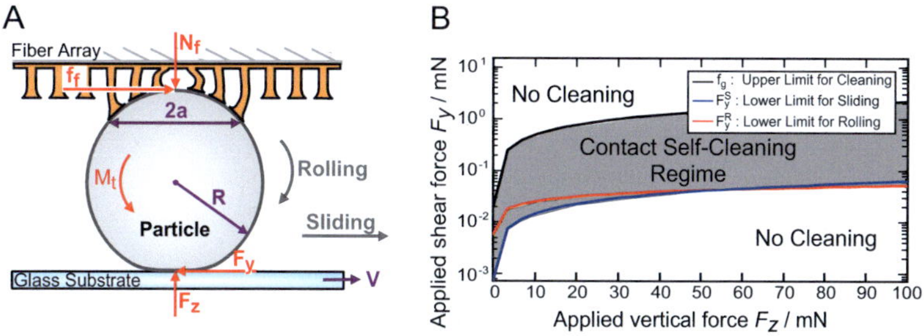

Figure 4.17.: Modelling of the contact self-cleaning mechanics in the large contaminants regime. **A** Schematic of the forces and moments affecting particle cleaning. F_y and F_z represent the applied shear force and the applied normal load, respectively; f_f is the friction force at the fiber-particle interface; f_g is the friction force at the substrate-particle interface; N_f is the normal force at the fiber-particle interface; M_t is the rolling resistance. **B** The theoretical cleaning condition (equation 4.6) reveals that cleaning by rolling or sliding can be achieved even for lowest load by choosing an appropriate combination of applied shear and normal forces. The shown plot corresponds to a fiber array (E_f = 2.9 MPa, ν_f = 0.49, and W_f = 93 mJ m^{-2}) contaminated with a 150 µm diameter glass sphere that is rubbed along a glass substrate (E_g = 73 GPa, ν_g = 0.17, and W_g = 56 mJ m^{-2}).

To *roll* the contaminant across the fiber array, its rolling resistance M_t must be overcome by the moment induced by the applied shear force. This leads to[123]

$$2F_yR \geq M_t = 6\pi R\left(W_f\xi_f + W_g\xi_g\right), \tag{4.3}$$

with W_f and W_g representing the work of adhesion at the fiber-contaminant and substrate-contaminant interfaces, respectively, and ξ_f and ξ_g are the shift in the contact area of the fiber-contaminant and substrate-contaminant interfaces due to rocking motion. Typically, the shift due to rocking motion is limited by the interatomic distance ε and the contact radius a, leading to

$$\varepsilon_f \leq \xi_f \leq a_f \quad \text{and} \quad \varepsilon_g \leq \xi_g \leq a_g. \tag{4.4}$$

In making the conservative assumption that ξ equals a, the shift in the contact area ξ becomes dependend on the applied normal load F_z according to eq. 4.3.

Finally, the static frictional force of the substrate/contaminant interface represents the upper boundary for the applied shear force F_y. Exceeding this static frictional force leads to *slipping* in the substrate-contaminant interface and thereby inhibition of lateral force transmission. Therefore, it is

$$F_y \leq \tau_g A_g, \tag{4.5}$$

where τ_g is the shear strength and A_g is the contact area of the substrate-contaminant interface.

Thus, the condition for cleaning a contaminating particle by rolling or sliding is given by

$$\tau_g A_g (F_z) > F_y \geq \begin{cases} \tau_f A_f (F_z) & \text{sliding} \\ 3\pi \left(W_f \xi_f (F_z) + W_g \xi_g (F_z)\right) & \text{rolling} \end{cases} \tag{4.6}$$

In (Figure 4.17B), the condition for cleaning a contaminating particle by rolling or sliding is plotted for typical values. Interestingly, even for lowest external loads there always exists a combination of applied shear and normal forces in which contact self-cleaning occurs.

4.5. Conclusion and Outlook

In summary, different sizes of soft mushroom-shaped microfibers were fabricated by soft molding and dipping processes. To reveal size effects in self-cleaning, these gecko-like adhesives were contaminated with spherical glass particles of different size ranges. In the adhesion and cleaning experiments it was observed that the size ratio of microfibers and contaminants strongly affects the rate and degree of contact self-cleaning. Furthermore, the mechanics of contact self-cleaning were investigated, showing that rolling end embedding is a major principle of self-cleaning of gecko-like adhesives. The most rapid cleaning can be achieved in case of microfibers that are much smaller than the contaminant. In this way, contaminants can easily roll off the patch. Exploiting these observations, a synthetic gecko-like adhesive was achieved, matching the attachment strength (140 kPa) and self-cleaning (up to 100 %) of geckos very closely. As demonstrated in a lamellae-inspired adhesive, hierarchy is required for contact self-cleaning of large area patches. The design guidelines discovered in this work could inspire new gecko-like adhesives that could ensure robust adhesion in both laboratory and real world conditions.

A shortened version of this chapter was submitted as the article "Staying Sticky: Contact Self-Cleaning of Gecko-Inspired Adhesives". Y. Mengüç and M. Röhrig (equally contributing co-authors), U. Abusomwan, H. Hölscher an M. Sitti. The examined microfibers were obtained from nanoGriptech LLC. The testing setup was designed by Y. Mengüç. Testing and analysis were supported by Y. Mengüç, H. Hölscher and M. Sitti. The modeling section was basically set up by U. Abusomwan.

5. Large-Area Replication of Gecko-Inspired Micro- and Nanostructures

Phenomena discovered in nature like the self-cleaning of gecko toes[12] and lotus leaves,[9] sensor functionality of butterfly wings,[124,125] and the adhesion of geckos[22] and insects[49] are promising for advanced applications as well as for consumer products in every day life. For instance, synthetic gecko-like adhesives may be applied in medical engineering,[119,126] pick-and-place systems[27,127] as well as for self-cleaning adhesive tapes for reclosable containers or office supplies.[12,81] As in the examples of lotus leaf, butterfly wings, and gecko toes, the remarkable features found in nature are often based on surface structuring. A hierachical formation of micro- and nano-pillars is found in most plant cuticles, in natural photonic crystals like the wings of the Morpho butterfly,[128] as well as in attachment devices of lizards and insects.[9,49] For instance, the adhesive toe pads of Tokay geckos (*Gekko gecko*) are covered by a hierarchical formation of high aspect ratio hair. These hair are called setae and are about 4 µm in diameter and 100 µm in length. The setae split up to five times and, finally, they branch into hundreds of tiny endings in the nanometer range. The high aspect ratio of the hierarchical hair decreases the effective elastic modulus by five orders of magnitude, leading to the highest compliance.[45] Due to this, geckos achieve very intimate contact to flat and even to relatively rough surfaces, enabling them to climb walls and ceilings with only the help of van-der-Waals interactions.[15] Additionally, as described in Chapter 3 and 4, hierarchy enhances adhesion due to energy dissipation and provides space to accommodate contaminants while maintaining self-cleaning.

5.1. Polymer Replication Processes

In order to mimic the surfaces found in nature, typically the established processes for the replication of micro- and nanostructures are applied. Injection molding is the most common technique and well established in industry. Recently, UV-nanoimprint increasingly gains significance due to its high precision. Hot embossing is the most universal micro replication technique making it particularly attractive for research facilities.

Injection molding. Injection molding is the method of choice for most macroscopic polymer components. A plastifying screw is doses, melts and compresses the polymer. Then, the low-viscous polymer melt is injected into a more or less complex mold. Typically, ejector pins embedded into the mold enable demolding. A broad variety of processable thermoplastic polymers and short cycle times makes injection molding economically interesting. Setting up temperature control of the mold makes this so-called variothermal injection molding process applicable for microstructures.[37] Molds heated above the softening temperature of the polymer prevent the polymer melt from solidifying within the small cavities. However, cycle times are increased due to the cyclic heating and cooling of the mold. Characteristically, long flow paths combined with high injection speed and rapid cooling of the polymer melt induce high inner stresses in the molded parts.[41] Additionally, the pressure drop occurring over the flow path leads to limited filling of exterior cavities and anisotropic shrinkage of the part.[39] These characteristics limit the applicability of variothermal injection molding for the fabrication of thin foils covered with high aspect ratio micro- and nano-structures.

UV-nanoimprint. UV-nanoimprint is a highly precise replication process for dimensions down to the nanometer range.[38] The micro- or nanostructured master, typically a PDMS or glass mold, is pressed into the uncured UV-curable polymer. UV-exposure initiates crosslinking and therefore curing of the polymer. Soft molds made of PDMS even allow for the demolding of undercuts. UV-nanoimprint is suitable for the large-area fabrication of micro- and nano-structures. However, it is restricted to a comparatively small variety of polymers. Additionally, demolding forces occurring from interfacial adhesion of the UV-curable polymer and the mold limit the achievable aspect ratio of micro- and nanostructures.

Hot embossing. A related technique is hot embossing[39] also known as thermal nanoimprint. The tremendous variety of applicable materials, ranging from amorphous and semicrystalline thermoplastic polymers to thermoplastic elastomers and to novel material classes like shape-memory polymers, liquid wood or metallic glasses, makes it the most universal micro- and nanoreplication technique.[39–41,129,130] In this open-tool technique, a thin polymer film is inserted between the mold insert and the opposing, typically rough, substrate plate (Figure 5.1). By heating up mold insert and substrate plate, the polymer is softened. Compression with forces up to 1000 kN ensures complete filling of the micro- and nanocavities. After solidifying the polymer by cooling the tool while maintaining the compression, the separation of mold insert and substrate plate demolds the delicate structures.[39] Here, cycle times are comparable to those of variothermal injection molding.[39] In contrast to injection molding, hot embossing excels in extremely short flow paths minimizing inner stresses induced in the molded parts. In addition, the viscousity of the polymer exceeds the one needed for injection molding. Consequently, a lower polymer temperature is sufficient leading to reduced shrinkage during cooling. Therefore, demolding forces are decreased allowing for higher aspect ratios of micro- and nano-structures compared to micro injection molding.[40] However, demolding forces still limit the aspect ratio of sub-micron structures to 3.[42]

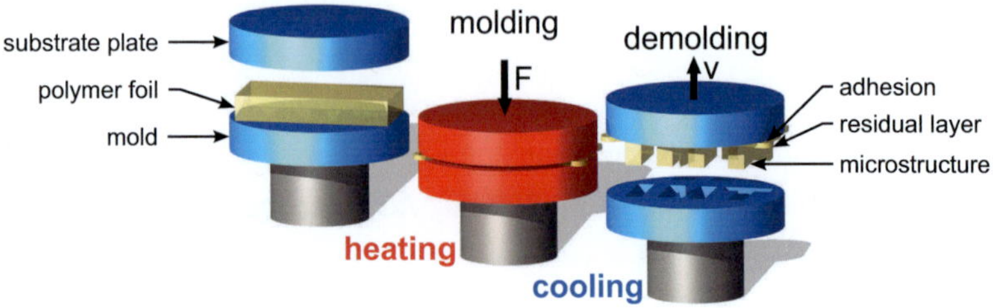

Figure 5.1.: Schematic of the hot embossing process.

Requirements in Biomimetics. Most bio-inspired surfaces, however, require not only higher aspect ratios but also the *hierarchical* formation of micro- and nanostructures. Today, limitations in aspect ratio, multilevel hierarchy and scalability of the established replication technologies impede commercial break-through of various bio-inspired smart surfaces. In order to bridge this gap, threefold hierarchical gecko-inspired micro- and nano-structures were fabricated by advanced hot embossing techniques.

5.2. Advanced Hot Embossing Techniques

In this section, advanced hot embossing and hot pulling processes allowing multilevel hierarchies of micro- and nanostructures with aspect ratio greater than 10 are presented. Threefold hierarchical gecko-inspired structures with 200 nm wide endings were fabricated. The feasibility of fabricating thin films covered with hierarchical micro- and nano-structures combined with cycles times comparable to the ones in variothermal injection molding makes these processes promising for the cost-effective fabrication of bio-inspired surfaces.

5.2.1. Hot Embossing

Hot embossing comprises of the softening and compression molding of a material into a mold insert. After solidification by cooling, the micro- or nanostructured sample is released from the mold insert by opening the tool (Figure 5.1 and 5.2A). In this way, the large-area replication of micro- and nanostructured parts with low inner stresses is feasible.

First, a foil of the chosen material is positioned between a rough substrate plate and the opposing mold insert containing the inverse copy of the master structure. The foil thickness has to exceed the depth of the master structure and is typically below 100 µm up to several millimeters.[41] Depending on the dimensions of the master structure, the metallic mold insert is fabricated by either micro-machining or the famous LiGA technique.[131] Recently, so-called shim mold inserts exhibiting thicknesses below 500 µm have attained more and more significance since fabrication is less time consuming compared to that of classical LiGA mold inserts reaching a few centimeters in thickness.[40] However, the decreased mechanical robustness of shim mold inserts is still a challenge in machine technology.[132] After positioning the foil, the tool is evacuated in order to allow complete filling of the cavities. Heating of mold insert and substrate plate ensures softening of the chosen material. The heating temperature ranges from below 90 °C in case of standard thermoplastic polymers to more than 450 °C for metallic glasses.[133]

In a velocity- and force controlled compression the highly viscous material is pressed into the mold insert. Typically, embossing velocities in the range of 1 - 10 $\mu m\,s^{-1}$ are chosen. The applied load ranges between 10 - 100 MPa and is controlled by a high precision load cell.[41] In order to ensure homogeneous filling of the mold insert, temperature and load are kept constant for a certain holding time. The complete process is controlled by a computer-assisted electronics that readjusts the temperature and position of thc tool.

In order to solidify the material, substrate plate and mold insert are cooled below the softening temperature of the material. During cooling the applied load is maintained to minimize shrinkage. Typically, shrinkage of the chosen material exceeds that of the metallic mold inserts. Hence, the replicated micro- and nanostructures press against the sidewalls of the cavities of the mold insert. Since this load may damage the replicated structures, demolding is the most critical step.[39] Consequently, demolding has to be controlled very precisely. After venting, separating the substrate plate from the mold insert opens the tool. Demolding of the replicated structures is based on the high adhesion of the replicated part and the rough substrate plate exceeding stiction of the filled mold insert.[41]

Experience in nano imprinting shows that even polymeric structures with dimensions below the size of their macromolecules are replicated very well. Consequently, it is considered that, up to a certain extent, the macromolecules adapt to the shape of the structured mold insert.[40] Hence, smallest dimensions are moldable in principle. However, the lateral dimensions and the achievable aspect ratio are limited by the quality of the mold insert and the occurring demolding forces. In the case of microstructures, aspect ratios as high as ten are considered ambitious.[40] As with decreasing structures the surface area increases in relation to the volume, nanostructures are especially sensitive to demolding forces. The smallest imperfection of the mold insert or minimal shrinkage may rupture the structures. Typically, these handicaps impede aspect ratios of more than 3 for hot embossing of nanopillars.[42] Therefore, below a certain size, hierarchical micro- and nanostructures are not moldable concurrently due to their high surface area and high sensitivity to demolding forces. In order to replicate these hierarchical micro- and nanostructures, hierarchical hot embossing and hot pulling processes are presented gradually enabling the fabrication of different levels.

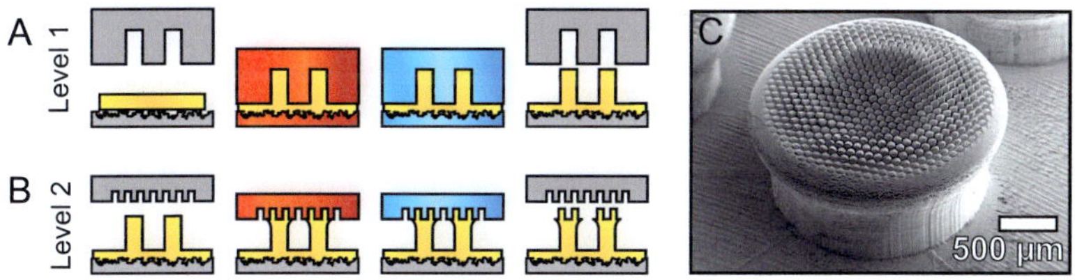

Figure 5.2.: A Hot Embossing can be divided into four major process steps. First, after inserting the polymer film between the micro- or nanostructured mold insert and the opposing substrate plate the tool is heated to the molding temperature. In the isothermal molding process the softened polymer is pressed into the mold insert's cavities. While maintaining the applied force, the polymer is cooled below its softening temperature. When solidified, the structures are demolded by opening the tool. **B** In order to add an additional level of micro- or nanostructures, the mold insert is replaced. In the hierarchical hot embossing step, only the mold insert is heated whereas the substrate plate remains at ambient temperature. When the hot mold insert exceeds the softening temperature of the polymer, it is moved towards the polymer structure until is touches the structure's top. The mold insert is pressed into the previous structure by a given distance or until a preset force is reached. The applied load presses the softened polymer into the delicate cavities and widens the endings of the previous layer. **C** The SEM image shows a twofold hierarchy (polymethylmethacrylate) reminiscent of the adhesive pads of tree frogs.[134] The secondary level honeycomb structures have a 40 µm edge length.

5.2.2. Hierarchical Hot Embossing

The hierachical formation of micro- and nano-pillars is essential for many of nature's smart surfaces.[9,10,22] In the case of gecko-inspired dry adhesives, hierarchy increases the adaptability to rough surfaces and contributes to the work of adhesion.[54] However, hot embossing of several hierarchical layers of micro- and nanostructures at once is unfeasible due to enormous demolding forces. Furthermore, hierarchy makes high demands of the flexibility of the mold fabrication process that are not met by classical techniques. Consequently, such hierarchical structures are replicated layer by layer in this work (Figure 5.2B). In order to add an additional micro- or nanostructured layer on top of existing structures the replaced mold insert is heated. After reaching the molding temperature, the mold insert is approached towards the structures until it touches their top. By applying a distinct load and penetrating into the underlying structures the softened material fills the mold insert's cavities. Additionally, the endings of the existing structures widen. After solidification by cooling the mold insert, the tool opens and demolds the fabricated layer. Figure 5.2C shows a scanning-electron-microscopy (SEM) image of a twofold hierarchical pillar inspired by the adhesive toes of tree frogs.[134] The endings of the basic layer are deformed resulting in mushroom-like tip shapes being advantageous for dry adhesives due to an increase in area fraction and improved mechanics.[33,55]

However, with decreasing dimensions, touching the structures becomes the crucial step. With established hot embossing machines, touching the structures cannot be sensed. The installed load cells are designed for typical molding forces of several hundreds of Kilonewtons and their force

Figure 5.3.: Scanning electron microscopy image of flattened microstructures. Without an appropriate electromechanical sensor, hierarchical hot embossing does not allow for touching microstructures sensitively. Since touching cannot be resolved by the force sensor, the microstructures are typically destroyed during approach.

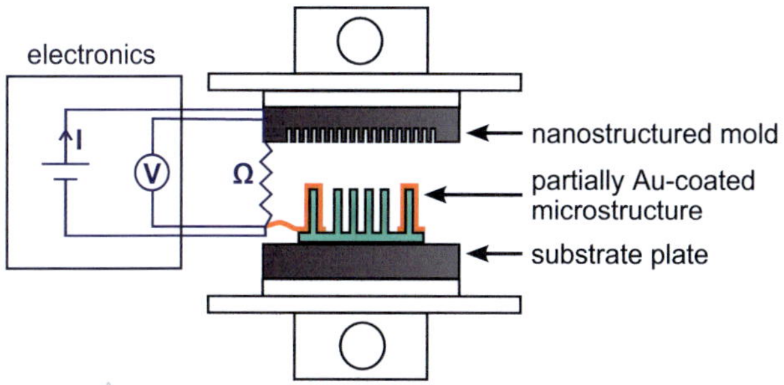

Figure 5.4.: Schematic of the electromechanical sensor principle used in the hierarchical hot embossing process. Exterior reference structures are covered with a thin gold layer in order to ensure conductivity. After connecting the reference structures and the mold insert to a measuring module, the resistance is measured constantly. A sharp drop in the resistance signal indicates contact of mold insert and reference structure.

resolution is very limited. Therefore, touching high aspect ratio pillars that are just a few microns in height is not detectable due to noise in the force signal. However, the highest precision is required in these dimensions otherwise the structures are flattened (Figure 5.3).

Sensors, however, are difficult to implement for hierarchical hot embossing. At reasonable expense, neither optical nor force based approaches are feasible. Temperatures of more than 450 °C result in unpredictable thermal expansion of the machine and the sample. Forces of up to 1000 kN destroy commonly implemented high resolution force sensors in the flux of force. In order to solve this challenge, electromechanical sensing was implemented into the hot embossing machines (Figure 5.4).

Electromechanical sensing is based upon measuring the electrical resistance between the metallic mold insert and separate reference structures. These reference structures are equivalent to the actual structures but are positioned outside the main structure field. While hot embossing of the basic layer a thin copper wire is embedded into the residual layer. Additionally, the reference structures are covered with a thin gold

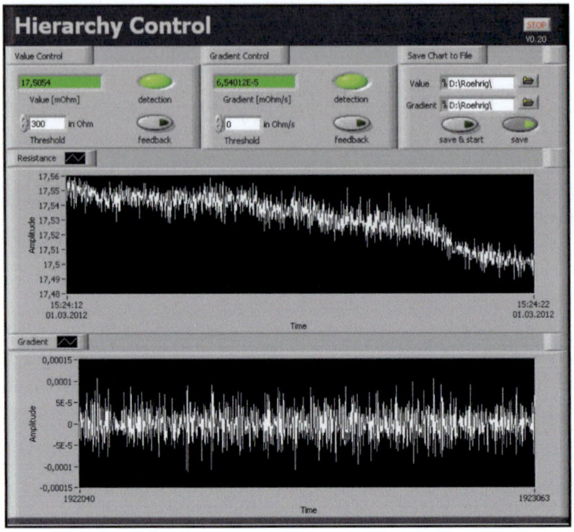

Figure 5.5.: Screenshot of the coded software (LabVIEW, National Instruments) controlling hierarchical hot embossing. The software accesses the commercial hot embossing software to trigger molding.

layer making them electrically conductive. Only these outer reference structures are covered with gold whereas the actual structure field remains unmodified. The sputtered gold layer (thickness: $\sim 50\,\text{nm}$) on the reference structures connects the structures' peak with the wire ensuring sufficient conductivity.[135,136] The mold insert and the embedded wire are connected to a measuring module. Based on the applied four-point probe resistance measurement,[137] the measurement signal is transmitted to a custom software interface (Figure 5.5). The software interfaces with the commercial hot embossing software. If the measured resistance drops below a defined setpoint by the mold insert touching the gold-covered structure, hierarchical embossing is triggered by the software. The reaction time of 3 seconds allows for touch accuracies below 500 nm achievable for touch-sensitive velocities (Figure 5.6). This represents a big steps towards reproducibility of hierarchical micro- and nanostructures (Figure 5.7).

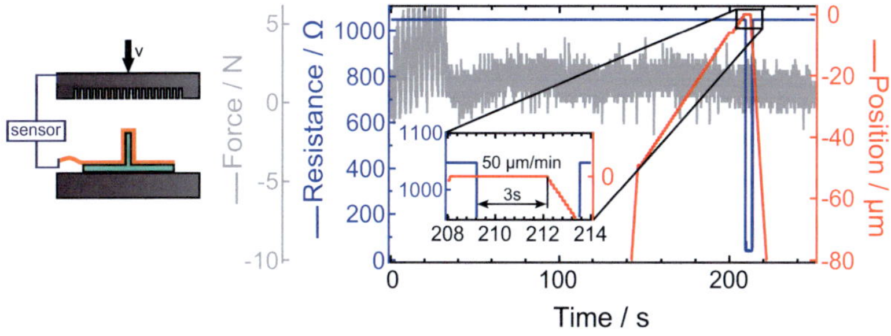

Figure 5.6.: Accuracy and reaction time of the electromechanical sensor. The plot shows the force, resistance and position signal while touching a single micro-pillar. Apparently, the force signal does not indicate the contact whereas the resistance signal sharply drops. The automated process requires 3 s reaction time in order to stop movement. This allows for touch accuracies below 500 nm for appropriate velocities.

Figure 5.8 shows the performance of the electromechanical sensor in a proof-of-principle experiment. First, micro-pillars of different heights were fabricated by hot embossing. A thin layer of gold was sputtered on top of the micro-pillars in order to ensure reliable conductivity. After connecting

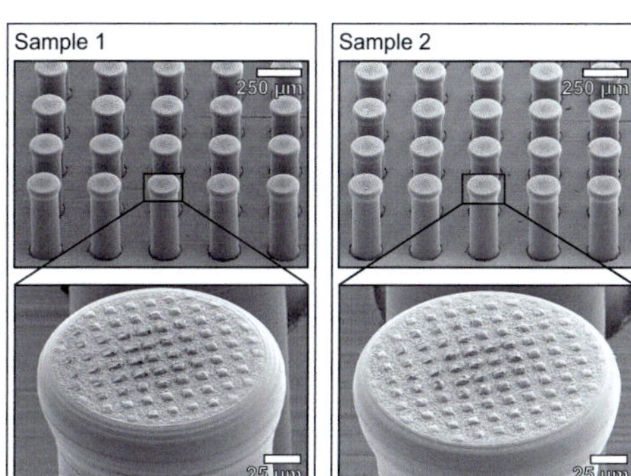

Figure 5.7.: Scanning electron microscopy images of two hierarchical samples. Both samples were fabricated in an automated hierarchical hot embossing processs using the electromechanical sensor. Both samples look identical showing the reproducibility of these twofold hierarchical samples.

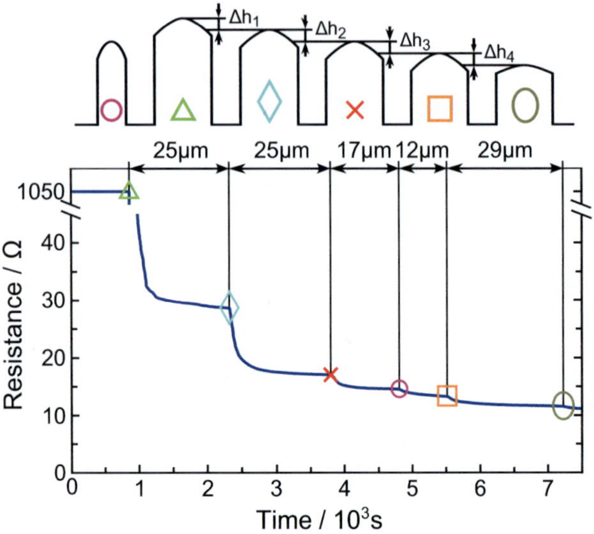

Figure 5.8.: In a proof-of-principle experiment, a mold insert was moved towards gold-coated reference pillars of different heights. Starting with the sensor's upper boundary value (1050 Ω), the resistance drops below 50 Ω when touching the highest pillar. The pillars below are sensed one after another as indicated by further drops in the resistance signal.

to the sensor, a shim mold insert was moved towards the micro-pillars at a constant velocity. The position and the measured resistance between the shim mold insert and micro-pillars were recorded. When touching the highest micro-pillar, the resistance dropped from the sensor's upper boundary value (1050 Ω) to less than 50 Ω (green triangle). The subsequent micro-pillars were sensed one after another indicated by drops in the resistance signal. The drops become smaller with increasing conductive area that is in contact with the mold insert.

Analyzing the resistance and position data even allows for the detection of height differences of the micro-pillars. In this way, the electromechanical sensor may also control tilting of mold insert and sample: for an array of micro-pillars with three exterior gold-covered micro-pillars for sensing, the resistance signal should drop at once in case of a perfectly aligned tool. Showing three distinct drops, however, indicates misalignment. A warning may provoke intervention by the operator in order to realign the tool.

5.2.3. Hot Pulling

The largest possible aspect ratios are required in order to increase compliance of gecko-inspired adhesives. In this way, the high aspect ratio hair covering the adhesive toes of geckos reduce their effective elastic modulus by five orders of magnitude compared to the bulk material properties.[45] However, the aspect ratio of replicated micro- and nanostructures is limited by the high demolding forces occurring during separation of the polymer and the mold insert. To overcome this problem an advanced hot embossing process is presented that even exploits these demolding forces in a pulling step in order to increase the aspect ratio. In order to distinguish this new molding technique from the well established hot embossing technique[39] it is called *hot pulling*. Applying hot pulling, nanopillars with aspect ratios of more than 10 were created successfully.

The hot pulling process differs from classical hot embossing mainly in the applied demolding temperature. Instead of solidifying the material before demolding by cooling the tool, the temperature of the mold insert is maintained above the material's softening temperature. While separating the mold insert, the viscous material elongates due to adhesion to the inner sidewalls of the cavities. Hence, roughness and even undercuts of the cavities of the mold insert are beneficial for high aspect ratios. Since demolding forces are even required in hot pulling, they can be increased by etching the sidewalls of the mold insert's cavities for example.

In contrast to the nanodrawing method reported by Jeong *et al.*[80] filling of the mold insert is not capillary driven. In the hot pulling process, pressing the mold insert into the polymer fills the cavities by a squeeze flow

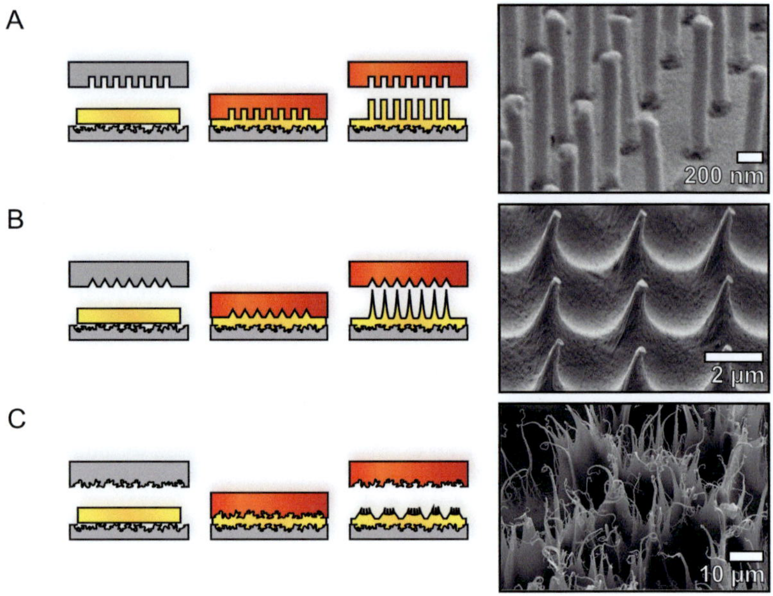

Figure 5.9.: In hot pulling the mold insert remains hot while demolding. The occurring demolding forces elongate the softened polymer leading to high aspect ratio micro- or nanostructures. **A** By using a LiGA-fabricated (electron beam lithography) shim mold insert, nano-pillars which are 140 nm wide and 1.4 µm in height were replicated. This clearly demonstrates the enormous aspect ratios achievable by hot pulling. **B** A wavelike structure, typical for interference lithography based LiGA mold inserts, could be elongated successfully. The SEM image shows spikes which are 270 nm wide at their tips and 3.2 µm in overall length. **C** Not only regular mold inserts but also mold inserts with stochastic topography can be used for hot pulling. By using a sandblasted steel-plate a promising nanofur consisting of dense high aspect ratio hair was fabricated. These hair are less than 200 nm in diameters and can be up to several microns long. *Polycarbonate was used for all samples shown in this figure.*

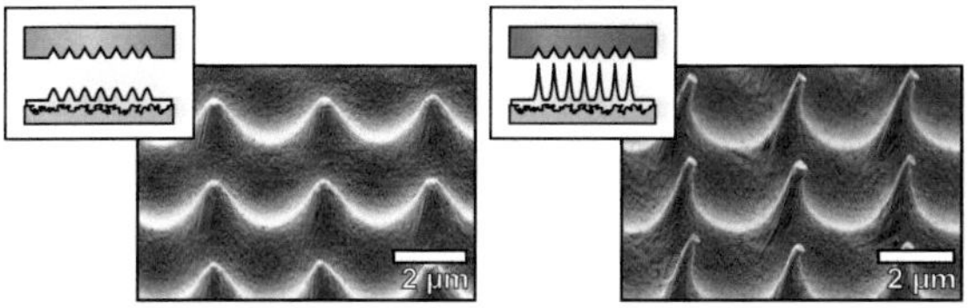

Figure 5.10.: Scanning electron microscopy images of two samples fabricated by hot embossing (left) and hot pulling (right), respectively. As can be seen, the aspect-ratio significantly increases as a result of hot pulling.

of the softened polymer.[39] In this way, well established semi-finished polymer sheets can be used instead of preparing an elaborate thin polymer film onto a silicon wafer. Furthermore, the polymer has not to be melted entirely ensuring short cycle times.

As shown in Figure 5.9, the hot pulling process is applicable for a wide variety of structures. Applying LiGA mold inserts based on electron beam lithography, freestanding nanopillars with aspect ratio of 10 were fabricated (Figure 5.9A).

In contrast to electron beam lithography, interference lithography is suitable for the large-area fabrication of wave-like patterns. With a corresponding mold insert, spikes with 270 nm wide tips and a height of 3.2 µm were achieved (Figure 5.9B and Figure 5.10 for comparison with hot embossing).

Both nanohairs and spikes required well-defined LiGA mold inserts. The fabrication of such mold inserts, however, is time-consuming and potentially very expensive. Sandblasting a steel-plate is probably one of the most cost-effective methods to create a large-area surface structure. As can be seen from Figure 5.9C, applying such a stochastic mold insert a promising nanofur was fabricated consisting of densely packed nanohair which are 200 nm in diameter and several microns high. In this way, the expensive fabrication of a well-defined mold insert can be avoided.

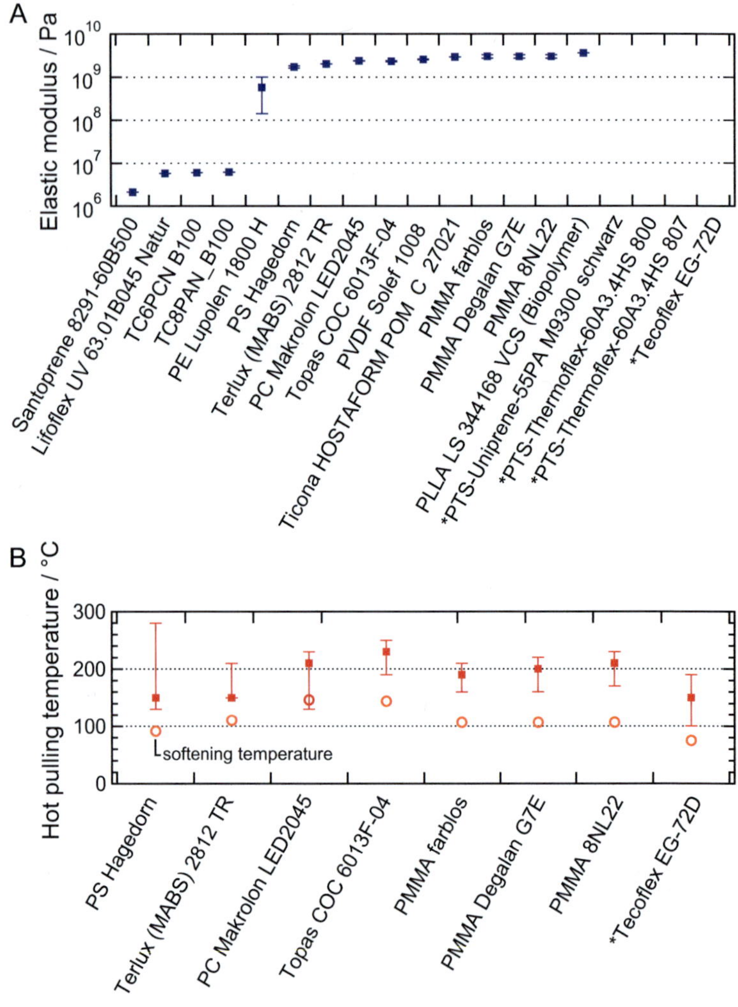

Figure 5.11.: **A** Elastic moduli of the chosen hot pulling materials. The materials were chosen in such a way that their elastic moduli cover a wide range. The elastic moduli were unknown for the materials marked with a *. **B** Demolding temperatures (quadrangles) for materials hot pulling works with. These temperatures are clearly above the softening temperature (circles) of the materials. The bars indicate the range of suitable demolding temperatures.

Besides geometry, material properties, especially the elastic modulus, affect the compliance of gecko-inspired dry adhesives. For a given geometry, compliance increases with decreasing elastic modulus. In order to identify materials suitable for the hot pulling of dry adhesives, several materials were tested to investigate their applicability for hot pulling. The elastic moduli of the selected materials range from 10^6 to 10^9 Pascal (Figure 5.11A). With sandblasted steel plates serving as mold inserts, hot pulling was performed with different parameter sets for each material according to Figure 5.9C. Particularly well-suited are the materials listed in Figure 5.11B.

Since the temperature of the mold insert is maintained constant for the complete process, cycle times are much lower compared to hot embossing, which requires time consuming heating and cooling cycles. Without optimization, cycle times below five seconds for the forming process are straightforward and indicate the cost-effectiveness of the hot pulling process.

5.3. Combination of Hierarchical Hot Embossing and Hot Pulling

In order to demonstrate the capability for the fabrication of bio-inspired smart surfaces, the developed processes were combined to fabricate a three-fold hierarchy inspired by the adhesive toes of geckos (Figure 5.12). With the first and second levels possessing lateral dimensions in the micron range, the third level ends with 200 nm thin hair. In this way, a density of 10^5 hairs per mm^2 was obtained.

First, with a classical hot embossing process an array of pillars (diameter: 450 µm, height: 1300 µm) was fabricated. After exchanging the mold insert, quadrangular pillars (edge length: 18 µm, height: 30 µm) were added in a hierarchical hot embossing step. Subsequently, a

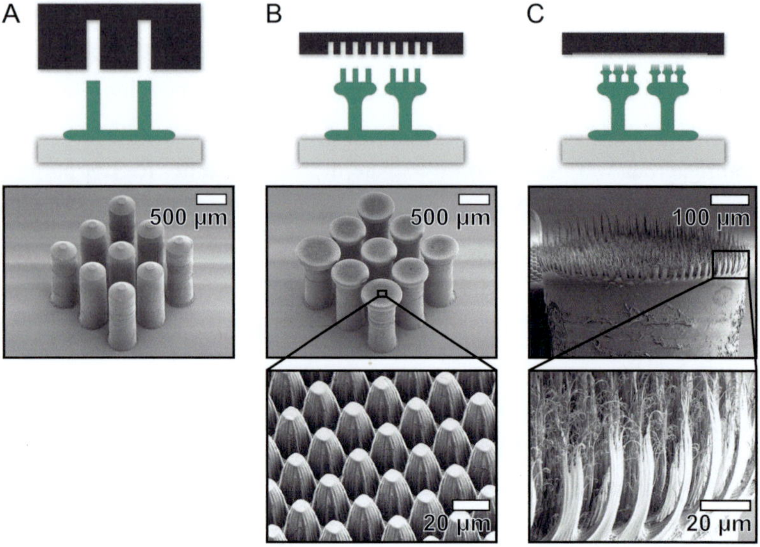

Figure 5.12.: Combining classical hot embossing, hierarchical hot embossing and hot pulling a threefold hierarchy of micro- and nanostructures was fabricated successfully (polycarbonate). **A** In a classical hot embossing process the lowermost pillars were replicated first. **B** With a second mold, quadrangular pillars were molded on top in a hierarchical hot embossing step. **C** Applying hot pulling, filigree high aspect ratio hair were pulled out with a nanostructured mold insert. In doing so, even the quadrangular pillars of the second level were enormously elongated.

hierarchical hot pulling process was performed with a third mold insert (period: 3 µm, depth: 5.4 µm). In doing so, the second level was elongated from 30 µm to ~50 µm. At its end the second level branches into high aspect ratio nano-hair (diameter: ~200 nm, height: up to 45 µm).

The limits of classical hot embossing are exceeded in this example. Multilevel hierarchy is achievable by subsequent fabrication of the particular levels. Maintaining the polymer in a softened state during demolding, highest aspect ratios of filigree micro- and nanostructures are pulled out of the polymer. In this way, the capability of the presented advanced hot embossing and hot pulling process is demonstrated.

5.3.1. Adhesion Measurements

In order to investigate the dry adhesion of the threefold hierarchical gecko-inspired structures, adhesion measurements were performed in accordance to the procedure described in section 3.2. Accordingly, a spherical silica particle (∅ ≈ 20 µm) mounted to a tipless cantilever (spring constant: $6.9\,\mathrm{N\,m^{-1}}$) was used as probe for the AFM adhesion measurements. In this way, force maps were plotted by ramping force-versus-distance measurements (distance: 7 µm).

Figure 5.13 exemplarily shows force-versus-distance measurements obtained on the threefold hierarchy and a flat polycarbonate foil as reference. After applying a preload of 100 nN, the threefold hierarchy exhibits adhesion of more than 400 nN during retraction (Figure 5.13A). Hence, the adhesion/preload ratio exceeds 4 for the gecko-inspired threefold hierarchy. Interestingly, the approach as well as the retraction curve show a lot of jumps probably occurring due to nanohairs that buckle/detach one after another. The enclosed area between both curves represents the work of adhesion required to separate the probe from the surface. Both curves do not merge until the start and end of the ramp at a distance of 7 µm away from the surface. This results in a comparatively large work of adhesion of 676 fJ. However, the force-versus-distance measurement looks curious: the measured force at start and end of the ramp (distance: 7 µm) is negative but should be 0 instead with the probe being retracted from the surface. The negative force at this point indicates that the high adhesion between threefold hierarchy and the spherical silica particle actually prevented getting the probe off the surface.

Figure 5.13B shows two different scales of a force-versus-distance measurement performed on a flat polycarbonate foil. In the magnified graph (top), the x-axis ranges from 0 µm to 0.5 µm distance. The graph reveals that less than 150 nN adhesion arises from a preload of 500 nN.

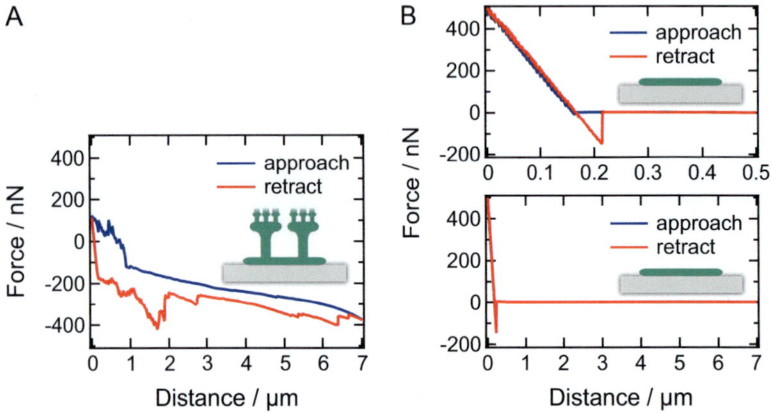

Figure 5.13.: Force-versus-distance measurements of the fabricated threefold hierarchy and a flat polycarbonate foil as reference. **A** The measurement of the threefold hierarchy shows, that after applying a preload of 100 nN adhesion of 400 nN occurs during retraction (adhesion / preload ratio: 4). The work of adhesion required for separation is 676 fJ. **B** Reference adhesion measurement of a flat polycarbonate foil. Both graphs show the same measurement, however, with different ranges of the x-axis. Applying a preload of 500 nN the spherical tip adheres with a force of less than 150 nN (adhesion/preload ratio: 0.3). As can be seen from the lower plot which is scaled identically to A, the enclosed area is considerably smaller meaning less work of adhesion is necessary to separate the probe from the surface (1 fJ).

Hence, the corresponding adhesion/preload ratio is less than 0.3 on the flat polycarbonate reference. The work of adhesion required to separate probe and surface was 1 fJ. The enormous difference in required work of adhesion compared to the threefold hierarchy is visualized when scaling both measurements equivalently (bottom).

Thus, the fabricated threefold hierarchy exhibits an adhesion/preload ratio improved by more than 13 and a required work of adhesion exceeding the flat polycarbonate reference by a factor of 676.

5.4. Technological Limitations

The large-area fabrication of hierarchical micro- and nanostructures is a technical challenge. To increase the performance of hierarchical gecko-inspired adhesives, even larger aspect ratios and enhanced density is required compared to the fabricated threefold hierarchy shown in section 5.3. Additionally, the moment of inertia of the micro- and nanostructures has to be decreased to enhance compliance. This follows with smaller and smaller cross-sectional areas of the delicate structures having decreasing dimensions. Hence, the demand on the precision and robustness of the mold inserts rises. In order to enable the large-area replication of micro- and nanostructures not only do the micro- and nanocavities have to exhibit the highest quality but also the shape tolerance of the mold inserts must be highly precise. Tilting or deflection of the mold inserts lead to uncorrectable misalignment in the large-area replication of hierarchical micro- and nanostructures.

In fact, nickel shim mold inserts typically exhibit a saucer-shaped distortion (Figure 5.14 A). Deflections of more than 300 µm can occur due to isothermal layer growth and thermally induced tension as a result of differing thermal expansion of the multi-layer composite while electroplating the mold insert.[132]

In order to reveal the technical limitations derived from misalignment, hierarchical hot embossing of extremely ambitious micro- and nanostructures was performed. Both levels on their own excel due to highest aspect ratios at smallest lateral dimensions achieved by hot pulling. Figure 5.15A shows SEM images of the fabricated micropillars serving as the first level (edge length: 7 µm, height: 45 µm, period: 14.5 µm). With an aspect ratio of 10, the nanostructures shown in Figure 5.15B (diameter: 140 nm, height: 1.4 µm, period: 580 nm) represent the very compliant second level. In a hierarchical hot embossing process, the second

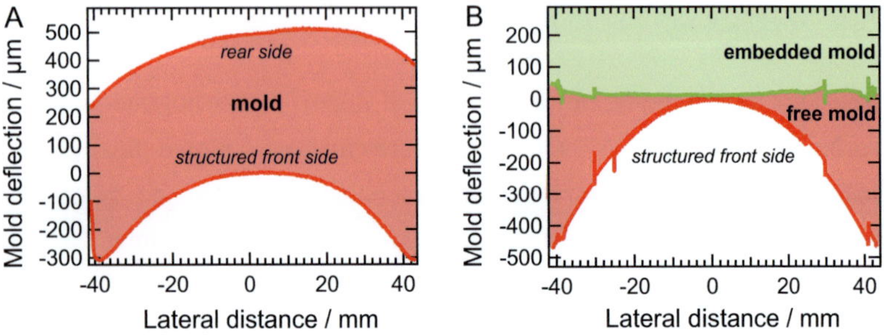

Figure 5.14.: A Deflection of a shim mold insert after electroplating (thickness: 500 µm, size: 4″) measured by laser triangulation (Werth VC 400HA). **B** Comparison of the distortion of a free shim mold insert before and an embedded mold insert after attaching to a polyamide 6.6 backing layer.

level nanostructures would be added on top of the first level microstructures. However, as can be seen from Figure 5.15C a precise large-area fabrication is prevented by misalignment. The first level structures at the left side of the SEM image are flattened whereas the structures on the right side are less compressed.

To overcome these issues in the future, novel fixing concepts have to be established enabling the recovery of the mold insert's deflection by embedding in an adhesive backing. First proof-of-principle experiments using adhesive backings of polyamide 6.6 are promising in terms of compensating for the distortion of shim mold inserts (Figure 5.14B).[132] Accordingly, fixing the mold inserts appropriately will provide engineers with the tools to replicate hierarchical micro-and nanostructures on large areas by advanced hot embossing processes.

5.5. Conclusion and Outlook

In conclusion, a combinatorial hot embossing and hot pulling process for the replication of multilevel hierarchical micro- and nanostructures was introduced. Fully automated, the introduced electromechanical sensor allows for the precise fabrication of hierarchies. In the presented hot pulling

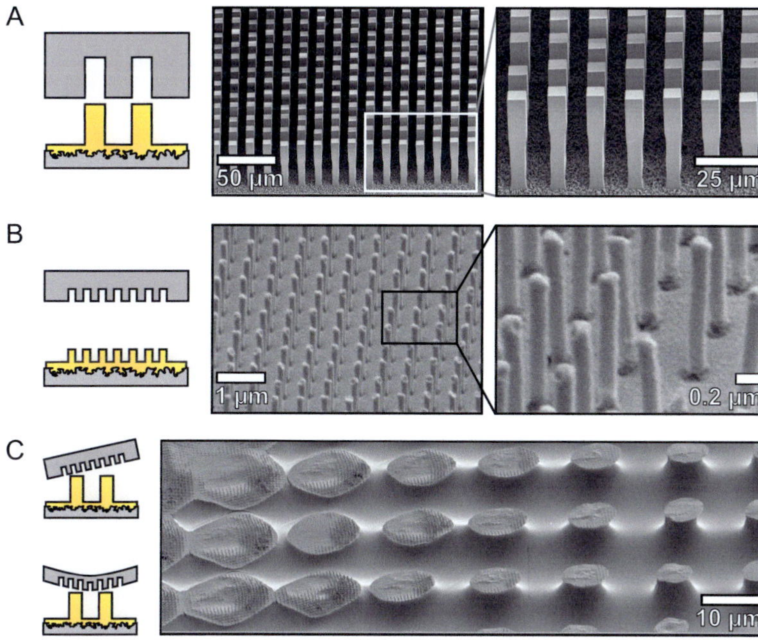

Figure 5.15.: Technological limitations of hierarchical hot embossing. **A** SEM images of the first level microstructures achieved by hot pulling. **B** The hot pulled nanostructures serving as second level exhibit aspect ratios of 10. **C** Combining the micro- and nanostructures in a hierarchical hot embossing step fails as a result of misalignment.

process, demolding forces that typically impede high aspect ratios of filigree structures are even utilized to elongate the polymer structures. In this way, an aspect ratio of ten was achieved for nanopillars which were 140 nm in diameter. Combining classical hot embossing, hierarchical hot embossing, and hot pulling gecko-inspired threefold hierarchical micro- and nanostructures were fabricated. 10^5 per mm^2 high aspect ratio fibers were achieved in this way. Compared to the flat reference, the threefold hierarchy improves adhesion and required work of adhesion by one and two orders of magnitude, respectively.

The enormous variety of applicable materials, the low inner stresses of the molded parts and the short cycle time make these processes economically viable. Scaling the processes up to role-to-role machines would open up the path to the mass production of hierarchical micro- and nanostructures with highest aspect ratios. In this way, nature can be mimicked cost-effectively for components with e.g. self-cleaning, sensor, or adhesive features.

A patent application was submitted that refers to electromechanical sensing used for measurement instrumentation and hierarchical hot embossing of micro- and nanostructures. M. Röhrig, M. Schneider, G. Etienne, M. Worgull and Hendrik Hölscher: Patent Application 102012110048.9 (2012).
A shortened version of this chapter was published as the article "Hot pulling and embossing of hierarchical nano- and micro-structures". M. Röhrig, M. Schneider, G. Etienne, F. Oulhadj, F. Pfannes, A. Kolew, M. Worgull and Hendrik Hölscher: Journal of Micromechanics and Microengineering ***23****, 105014 (2013).*

6. Hot Pulling of Nanofur for Biomimetic Applications

Nature designs fascinating surfaces to fulfill various purposes. Micro- and nanoscale structures play a key role in most of these surface effects. Two famous examples of nature's inventions are *gecko* toes and *lotus* leaves. Hierarchical hairs enable geckos to cling to nearly any surface by van-der-Waals forces[15] and microstructures covered with a fur of wax promote self-cleaning of lotus leaves.[9] The leaves of the lotus[8] and the floating water fern *salvinia*[10] provide specific wettability to ensure unobstructed photosynthesis by self-cleaning or by retaining an air-film when submerged underwater. And although *rose* petals are superhydrophobic, water droplets remain stuck to them even when hanging upside-down.[138] These effects are of interest not only for fundamental research[18, 139, 140] but also for technical applications such as self-cleaning windshields and antennas,[23] solar panels,[141] anti-fouling and drag reducing coatings[10] as well as energy conservation.[31]

Inspired by these and other examples, scientists and industry around the world are exploring new methods for the fabrication of such surfaces.[31, 33, 35, 80, 83, 84, 113, 141–145] However, a scalable and cost-effective method for mimicking nature's designs is still a challenge.

Here, a highly scalable molding technique for the fabrication of high aspect ratio nano-hair on polymer surfaces is introduced. These nano-hair surfaces are suitable for various biomimetic applications. In this presented hot pulling process, softened polymer is melt drawn with a heated sandblasted steel plate serving as mold insert, resulting in densely packed hairs that are pulled out of a polymer foil. The usage of a sandblasted steel plate

Figure 6.1.: **A** Droplets resting on the superhydrophobic nanofur form a spherical shape. **B** Due to its superhydrophobicity, only a minimal area of nanofur is required in order to carry a large volume of water ($\sim 1.25\,\mathrm{l\,m^{-2}}$).

as a mold insert guarantees minimized tooling costs and makes hot pulling a cost-effective fabrication method.

Conventionally, wetting of technical surfaces is controlled by chemical modification or by adding a defined micro- or nano-structure. Various techniques such as surface treatment with low surface energy compounds,[146,147] growing nanotubules[31,148] and molding techniques[80,144,147,149] are applied to fabricate water repellent surfaces. However, water repellency of the chemically modified but unstructured surfaces is strictly limited,[150] and superhydrophobicity requires specificly structured surfaces. Industrial implementation of surface structuring techniques, however, is impeded by expensive feedstocks, complex processes, necessity of sophisticated mold inserts or limited scalability.[147,151,152]

The hot pulling technique presented here on the other hand is able to switch hydrophilicity into superhydrophobicity (Figure 6.1) by a very easy and cost-effective structuring method. Sandblasted steel plates serve as mold inserts for fabricating superhydrophobic surfaces from polycarbonate foils. Without any chemical surface treatment contact angles on polycarbonate increase from 72° to 174°. Local physical damage of the surface topography returns the hydrophilicity of the polycarbonate, and since the surrounding nanofur is still superhydrophobic this feature opens a path to various applications in microfluidics. Liquid traps as well as high and low water adhesion are presented enabling the easy fabrication of microreactors. Additionally, the ability to retain air films underwater make these surfaces promising for drag reduction. Surface-treatment with low surface energy compounds (SLIPS[153]) makes them 'self-healing'. Further, superhydrophobicity combined with superoleophilicity makes the nanofur applicable for oil/water separation and filtration.

6.1. Short Introduction into Wetting

A sessile liquid droplet resting on a solid may form a spherical shape, slightly deformed by gravity. On the other hand, the droplet may spread on a solid and wet its surface. In this case, the contact angle θ at which the droplet joins the solid is comparatively small. Generally, for contact angles below 90° ($\theta < 90°$) the liquid wets the surface (hydrophilicity), whereas contact angles of more than 90° ($\theta \geq 90°$) refer to non-wettability (hydrophobicity).

Typically, molecules existing inside a liquid droplet feel intermolecular forces uniformly distributed in all spatial directions. However, molecules in the outer film of a droplet predominantly feel the interactions to the inside. Consequently, work is necessary in order to shift a molecule reversibly from inside the droplet to the less strongly bound outer film.[154] The work needed per unit surface is defined as the specific surface tension. However, all molecules in the outer film additionally interact with their environment. Hence, the term needs to be extended to the specific interfacial surface tension γ including the interactions with the surrounding molecules.

In 1805, T. Young[155] already described wettability quantitatively by connecting the contact angle and the specific interfacial surface tensions acting on the triple line of liquid, solid and vapor (in the following these phases are indicated by the indices L, S and V; see Figure 6.2 A)

$$\cos\theta = \frac{\gamma_{SV} - \gamma_{SL}}{\gamma_{LV}}. \quad (6.1)$$

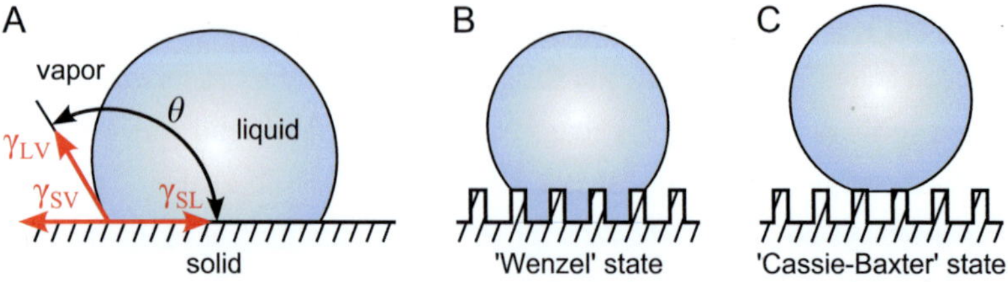

Figure 6.2.: Schematic of wetting states. **A** The contact angle θ at which the droplet joins the triple line of liquid, solid and vapor is defined by Young's equation. At the triple line, all interfacial surface tensions are in equilibrium. **B** A droplet resting on a structured surface may either penetrate into the surface topography ('Wenzel' state) or **C** the droplet may sit on top of the topography enclosing air between the structures ('Cassie-Baxter' state).

Wetting states. Conventionally, chemical modifications like fluorination are applied in order to enhance the hydrophobicity ($\theta \geq 90°$) of solids. However, even on extreme hydrophobic solids the contact angle won't exceed values higher than $\sim 120°$ if the surface is smooth (e.g. monomolecular CF_3).[150] Achieving higher contact angles requires a surface topography, e.g. roughness or a regularly structured surface.[11]

In general, surface topography leads to different wetting behavior (Figure 6.2B and C). First, the liquid may penetrate the surface topography. Introducing the roughness factor φ as ratio of the actual area of a surface topography to the projected area, Wenzel[156] determined the correlation

$$\cos\theta_W = \varphi \cos\theta_0, \tag{6.2}$$

with θ_W being the contact angle of the droplet penetrated into the topography and θ_0 representing the contact angle of the smooth solid. Consequently, a hydrophilic solid with surface topography should appear more hydrophilic than a smooth surface of the same material. The converse is also true: if a hydrophobic material has a surface topography, it should behave more hydrophobic than without.[18,157,158]

Second, instead of penetrating into the topography the droplet may partially rest on air pockets which are trapped within the topography. Under this condition, the Cassie-Baxter model[159] describes the contact angle of these so called 'fakir droplets':

$$\cos\theta_C = -1 + \Phi_{SL}(\varphi_\Phi cos\theta_0 + 1), \tag{6.3}$$

where θ_C represents the contact angle of a droplet in the Cassie-Baxter state and Φ_{SL} the fraction of the wetted solid. The roughness factor of the area wet by liquid is depicted by φ_Φ. Surfaces causing the Cassie-Baxter state are often superhydrophobic.

Contact angle hysteresis. High contact angles do not necessarily imply that droplets may roll off the surface easily. Chen *et al.*[160] presented superhydrophobic surfaces ($\theta = 169°$) on which droplets stuck even when hung upside down. Consequently, in order to characterize the mobility of droplets on surfaces, the authors emphasized to use another physical quantity: the contact angle hysteresis. Surface topography and other surface heterogeneities, e.g. in surface chemistry or simple contamination, allow the droplet to pin on these defects. As a result, the observed contact angle may have multiple values.[158] This phenomenon can be observed on a droplet resting on or rolling off a tilted plane. Typically, the contact angle at its front (advancing contact angle θ_a) is significantly higher than the observed contact angle at its rear (receding contact angle θ_r). The difference in the observed contact angles at front and rear causes capillary forces, that counteract gravity and can make droplets stick.[161] The difference between advancing and receding contact angle defines the contact angle hysteresis. Typically, a droplet in the Cassie-Baxter state offers significantly reduced contact angle hysteresis compared to Wenzel droplets, making them roll off much more easily.[162] This is a result of the trapped air layer covering the imperfections of the surface.[157]

6.2. Fabrication of Nanofur by Hot Pulling

Polymer components have to be produced cost-effectively to be successful in the mass market. For conventional polymer replication processes such as injection molding, thermoforming and extrusion, limits for the fabrication of micro- and nanostructured parts already come into sight.[163] Novel techniques such as UV nanoimprint and hot embossing excels for the replication of micro- and nanostructures.[39,164] However, as discussed in Chapter 5 the minimum achievable cross section and the maximum aspect ratio of the replicated structures are strictly limited by the friction forces arising during demolding.[39] Often the fabrication of high aspect ratio micro- and nanostructures even requires the destruction of the mold insert.[43,113,165]

Fortunately, the presented hot pulling process excludes expensive mold inserts. In this process, the occurring demolding forces are even utilized to form high aspect ratio micro- and nanostructures. A sandblasted steel-plate serves as the mold insert in the fully automated process (Figure 6.3A and B). After evacuation of the vacuum chamber, the mold insert is heated to a temperature exceeding the glass transition temperature of the material (for polycarbonate $T_g = 144\,°\mathrm{C}$). When this pre-defined temperature is attained, the mold insert is approached towards the opposing polycarbonate foil which is attached to the machine. As soon as the mold insert is in contact with the polycarbonate, softening of the polymer begins. Once the mold insert reaches the chosen depth of penetration and the predefined holding-time has elapsed, the heated mold insert is retracted from the polymer with a controlled velocity. Therefore, during this major step of hot pulling the softened polymer elongates during demolding and formes a cratered surface covered with high aspect ratio nanohairs (Figure 6.3C). After cooling and venting of the vacuum chamber, the finished polycarbonate sample is taken out of the opened machine. Further experimental details are specified in Appendix A.1.

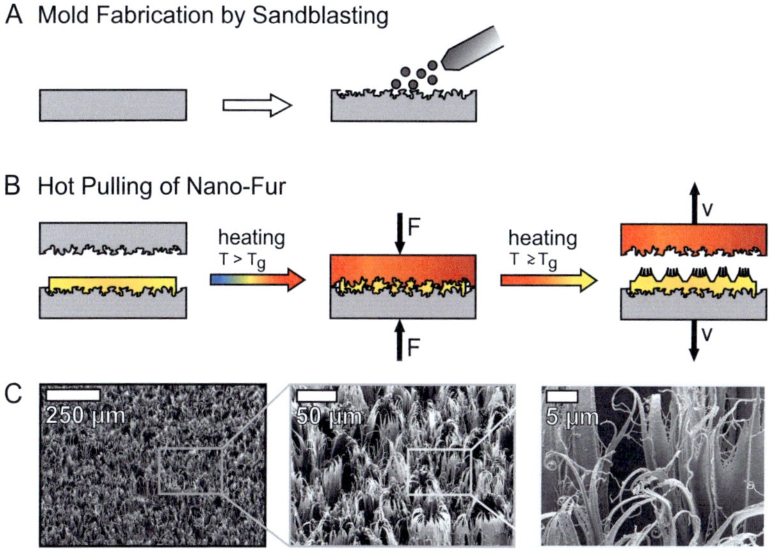

Figure 6.3.: Fabrication of nanofur by hot pulling. **A** A steel-plate is used as mold insert. The required roughness and undercuts are generated by sandblasting. **B** The heated mold insert is pressed into the foil in a fully automated process. The polymer softens and fills the cavities of the mold insert. By retracting the heated mold insert, the softened polymer elongates and forms a cratered surface covered with high aspect ratio nanohairs. **C** SEM micrographs showing the fractal surface topography of samples fabricated from polycarbonate.

In contrast to the hot embossing process, demolding in hot pulling occurs with a hot mold insert. In classical hot embossing the mold insert is cooled *below* the glass transition temperature of the polymer before demolding. However, the different shrinkage rates between the mold insert and the polymer results in higher demolding forces that frequently damage the micro- and nanostructures during demolding. In the hot pulling process, on the other hand, the mold insert is *not* cooled. Hence, shrinkage is not contributing to demolding forces. Instead, adhesion and normal forces at the undercuts of the sandblasted mold insert elongate the softened polymer. In this way, cratered surfaces with nanohairs covering the upper edges are created (Figure 6.3C). The fabricated nanohairs can be up to several microns long and less than 200 nm in diameter. Depending on the parameters, the diameter of the craters varies from 30 µm to 125 µm.

As far as can be ascertained, no other large-scale technique is able to produce nanofur in such an easy way. The nanodrawing technique reported by Jeong *et al.*[80] requires elaborate polymer thin films and a sophisticated mold. Applying 2½ dimensional lithographic methods is not suitable to mimick the arrangement of the nanofur. Additionally, high aspect ratio structures like the nanofur would probably tend to collapse during development. Furthermore, conventional molding techniques fail to produce such high aspect ratio nanostructures without destroying the mold insert due to high demolding forces.[43,113,165]

With a diameter of up to 8 inches for the processable area, already conventional hot embossing machines allow for the large-area application of the presented hot pulling process. Even with an unoptimized four second process duration, the cost-effectiveness of hot pulling is comparable to variotherm injection molding.[39] Adopting hot pulling to roll-to-roll embossing will open a route to even further cost-effective uninterrupted fabrication.[166,167]

6.3. Applications

6.3.1. Superhydrophobic Surfaces

In order to fabricate superhydrophobic surfaces ($\theta \geq 150°$[23]), chemical surface treatment alone is insufficient. In case of a smooth surface, the contact angle won't exceed $\sim 120°$ even for extreme hydrophobic substrates (e.g. monomolecular CF_3).[150] Consequently, superhydrophobicity requires micro- and nanostructured surfaces.

Applying the presented hot pulling process the contact angle of polycarbonate was increased by structuring from 72° up to 174°. In Figure 6.4 a contact angle measurement on a flat polycarbonate foil (left) and on the nanofur (right) are shown for comparison. For these measurements, 4.5 µl of deionized water was dispensed and analyzed with a commercial contact angle measuring system (OCA 40, DataPhysics Instruments; Appendix A.2). The flat polycarbonate foil exhibited a static contact angle of $72 \pm 4°$ (N = 10 measurements). Hence, the polycarbonate foil was hydrophilic ($\theta < 90°$). In contrast, the structured polycarbonate sample increased the static contact angle to $174 \pm 4°$ (N = 7 measurements) which is superhydrophobic ($\theta \geq 150°$[23]). The exceptionally high contact angles are mainly addressed to the fractal nature of the surface which is fully covered by tiny hairs.[157,168,169] A chemical modification of the surface did not occur by hot pulling , as confirmed by 'X-ray Photoelectron Spectroscopy' (XPS).

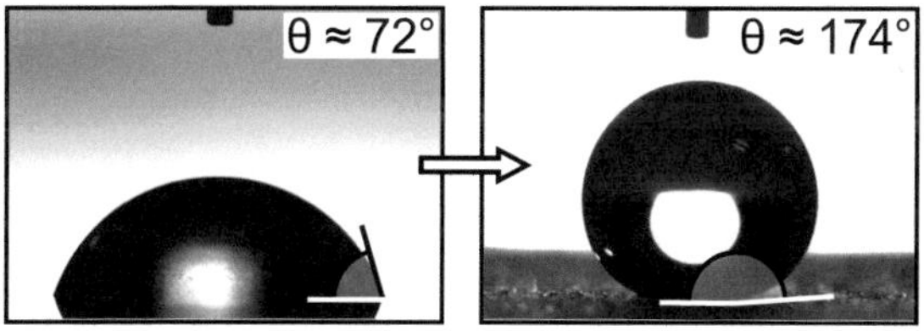

Figure 6.4.: Measurement of the static contact angle of a 4.5 µl water droplet on a flat polycarbonate foil (left) and on a polycarbonate nanofur sample fabricated by hot pulling (right). The fractal nature of the hot pulled sample leads to an increase in the static contact angle from 72° to 174°.

Droplet adhesion. By choosing the hot pulling parameters appropriately, the density of the nanofur can be varied (Table 6.1). Both nanohairs of low density (crater width ≈ 125 µm, Figure 6.6B) as well as very dense nanohairs (crater width ≈ 30 µm, Figure 6.6C) were fabricated. Different droplet adhesion was shown in an evaporation experiment (Figure 6.5A). For this purpose, a 4.5 µl droplet of deionized water was dispensed on samples with lower and higher density of nanohairs. In the beginning, both droplets exhibited a hydrophobic conctact angle. While evaporating, droplets typically reveal their receding contact angle.[170] After 50 minutes, the droplet on the high density sample (left sample in Figure 6.5A) still showed a high contact angle, indicating the Cassie-Baxter state.[159] The underlying air film in this state allows the evaporating droplets to change their contact line easily. However, the droplet on the lower density sample (right sample in Figure 6.5A) hardly changed its contact line while evaporating. The receding contact angle after 50 minutes was hydrophilic. This observation indicates a pinning droplet and a large contact angle hysteresis typically appearing in the Wenzel state, in which liquid penetrates the surface topography.[156]

A motion experiment on this 'pinning sample' is shown in Figure 6.5B where the droplet was pulled along the surface (left image) and retracted from the sample (right image) respectively. Significant pinning effects counteracting the motion are clearly visible.

Both high and low adhesion of droplets are advantageous for different applications. For example, pinning of droplets is a useful feature for avoiding water deposition (aircraft cabins, ceilings of clubs),[171] or holding liquids at specific positions (microfluidic devices).[158] On the other hand, low adhesion enables self-cleaning surfaces as exemplified by the lotus leaf[9] and positively affects fluid flow in microfluidic devices.[172,173]

Table 6.1.: Hot pulling parameters of the sliding sample (high density nanofur) and the pinning sample (low density nanofur).

	sliding sample	pinning sample
embossing temperature / °C	215	220
embossing velocity / mm min^{-1}	0.4	0.4
depth of penetration / µm	200	200
demolding temperature / °C	215	220
demolding velocity / mm min^{-1}	0.3	1.5

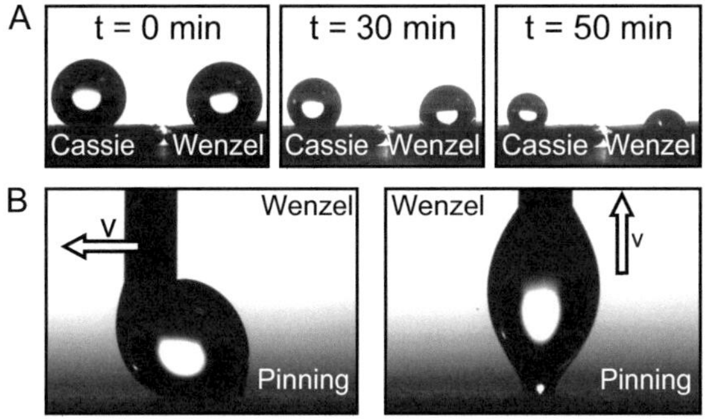

Figure 6.5.: **A** An evaporation experiment proves the wetting state on two different samples. As a function of the chosen parameter set, hot pulling leads to surfaces with different densities of nano-hairs. The sample on the left exhibits very dense nano-hairs ($\approx 150\,000\,\mathrm{mm}^{-2}$). Droplets dispensed on this surface stay in the Cassie-Baxter state. Consequently, the shrinking droplet is able to shift its contact line easily. Thus, the receding angle after 50 minutes is still hydrophobic. The sample on the right is covered with nano-hairs of lower density ($\approx 90\,000\,\mathrm{mm}^{-2}$). While evaporating, the contact area to the solid changes only marginally, resulting in a hydrophilic contact angle after 50 minutes. This observation indicates pinning effects at the surface leading to a high contact angle hysteresis, which is typical for droplets in the Wenzel state. **B** Motion experiments performed on the 'pinning sample' with a low density of nanohairs. The droplet was pulled along the surface (left) as well as retracted from the sample (right). It can be easily seen, that pinning effects occurred counteracting the pulling.

Droplet mobility. In order to investigate the spectrum of possible applications, the mobility of the droplets on two types of nanofur and a flat polycarbonate foil was compared. In tilting experiments, a droplet is dispensed onto the particular sample which is attached to a tilting stage which itself is adjusted horizontally. Afterwards, the stage is tilted until the droplet starts to move under the influence of gravity. The sliding angle α_s at which movement of the droplet occurrs is recorded.[174] The sliding angle characterizes how easily the droplet rolls off a surface and is thus decisive for the field of application.

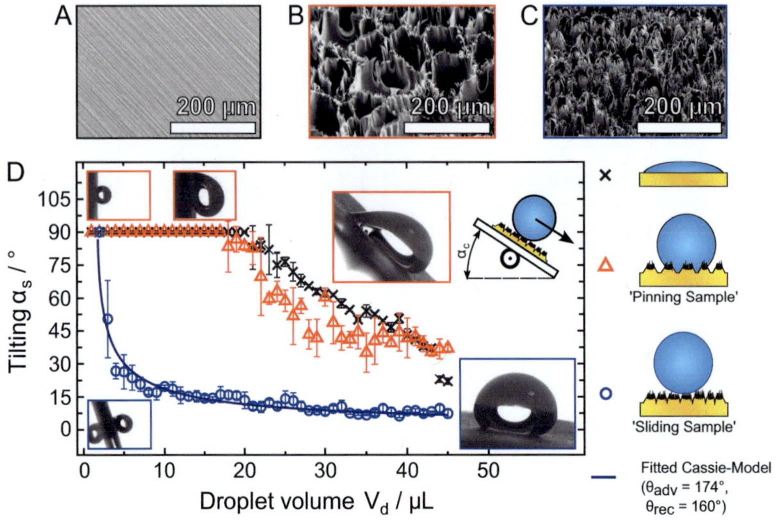

Figure 6.6.: Sliding angle measurement of water droplets on **A** a flat polycarbonate foil, **B** a 'pinning sample' and **C** a 'sliding sample'. The 'pinning'- as well as the 'sliding sample' were both fabricated by hot pulling a polycarbonate foil but with a different set of parameters. The hairs covering the surface of the 'pinning sample' are less dense compared to the ones covering the 'sliding sample'. **D** The measurement of the sliding angle was performed by attaching the samples to a tilting stage. After dispensing a droplet, the stage was tilted until the droplet started to slide, defining the sliding angle for the chosen volume of the droplet. The experiments were performed with droplet volumes ranging from 1 μl to 45 μl. Each measurement was repeated three times. Both the flat polycarbonate foil and the 'pinning sample' provide similar adhesion to the droplet. Droplets remain stuck for volumes up to 20 μl even when hanging upside down. With increasing volumes, the sliding angle falls nearly linearly in both cases. In contrast, droplets roll off the 'sliding sample' easily. Here, the underlying air film covers the surface heterogeneity minimizing pinning.

In Figure 6.6D, the results of the tilting experiments are shown. On the hydrophilic flat polycarbonate foil droplets with volumes of up to 20 μl stuck even when they hung upside-down (black crosses in Figure 6.6D). Then, starting with 90° the sliding angle decreases almost linearly with 2.3 °/μl. The advancing and receding contact angle of the flat polycarbonate foil are θ_a = 82.6° and θ_r = 36.6° respectively. Therefore, the contact angle hysteresis of the flat polycarbonate foil is θ_Δ = 46°.

Interestingly, the droplets dispensed on the 'pinning sample' remained stuck in a similar way. The advancing contact angle is very large (θ_a = 164° ±7°), however, the receding angle is very low (θ_a = 50° ±9°). The resulting contact angle hysteresis of θ_Δ = 114° ±13° is even greater than that measured on relevant rose petals (*Rosa Hybrid Tea, cv. Bairage* $\theta_\Delta \approx 90°$[175]), and is a clear evidence of high adhesion on the 'pinning samples'. Due to the surface topography an energy barrier has to be overcome in order to shift the contact line.[172,176] In addition to the capillary effects at the rear[161] this leads to high adhesion of the droplet, despite its high static contact angle (θ_W = 171° ±4°).

In comparison to the previously described observations, droplets dispensed onto the sample with dense nanohairs rolled off easily. Droplets with a volume of more than 20 µl rolled off provided that the 'sliding sample' is tilted at approximately 10°. The measured contact angles are θ_a = 167° ±3° for the advancing, and θ_a = 137° ±8° for the receding contact angle. Therefore, the resulting contact angle hysteresis (θ_Δ = 30° ±9°) is much less than the ones previously observed. In order to describe the condition for a fakir-droplet remaining stuck, Quéré *et al.*[161] provided an expression for calculating the contact angle hysteresis necessary for a droplet to adhere to the substrate at a distinct tilting angle. Rearranging equation (5) in reference[161] leads to

$$\alpha_s = arcsin\left(\frac{\theta_\Delta \sin^2(\theta_m)}{\frac{4^{2/3}}{3}(R\kappa)^2\,(2+\cos(\theta_m))^{1/3}\,(1-\cos(\theta_m))^{2/3}}\right), \qquad (6.4)$$

with $\theta_\Delta = \theta_a - \theta_r$ being the contact angle hysteresis, $\theta_m = (\theta_a + \theta_r)/2$ representing the average contact angle, $R = (3V/4\pi)^{1/3}$ is the radius of the droplet before it hits the solid, and $\kappa = 2.7\,\text{mm}$ is the capillary length for pure water. Fitting this equation to the sliding angle measurement led to $\theta_a = 174°$ for the advancing, and $\theta_a = 160°$ for the receding contact angles. The values resulting from this fit are close to our measurements.

Microfluidic applications. With the remarkable characteristic of hydrophilic polycarbonate transformed to a superhydrophobic surface by hot pulling, various applications of the nanofur are imaginable (Figure 6.7).

In order to confine a liquid on a 'sliding sample', location and shape of the droplet trap were easily defined by physically destroying the surface topography in the desired contour. In the experiment shown in Figure 6.7A, a stamp with a triangular contour was pressed into a 'sliding sample'. Applying a droplet, the contact line of the droplet immediately joined the triangular contour. Dispensing more water did not result in breaking through this pressed-in contour. The hydrophilic polycarbonate bared in the contour instead pinned the water to such an extent, that even tilting of the sample was possible without loosing the confined droplet.

Adhering the droplets in damaged regions can be also used to capture water. Figure 6.7B demonstrates that damaging the surface topography of a 'sliding sample' reveals the hydrophilic character of the polycarbonate. Water running across the sample was easily flowing over the structured area whereas it stuck to the damaged zones. In this way, arbitrary reservoirs can be designed with minimal effort.

Additionally, the presented surfaces are perfect substrates for chemical or biological liquid experiments. Smallest amounts of liquid can be easily manipulated without any residues. As a demonstration, Figure 6.7C shows the preparation of 'Turnbull's Blue' by mixing the aqueous solutions of 'tripotassium iron (3+) hexacyanide' and 'dichloroiron'.

Combined with the other features shown in Figure 6.7, the nanofur surfaces are applicable for a wide variety of applications in microfluidics.

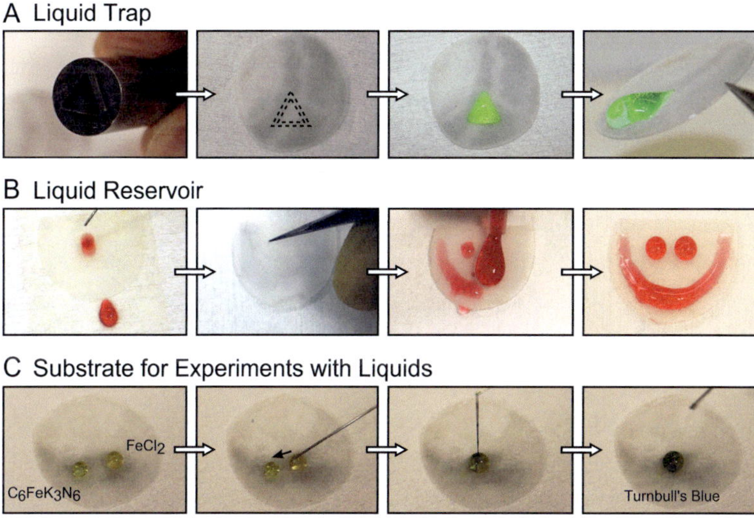

Figure 6.7.: Microfluidic applications of the nanofur: **A** By damaging the surface of the nanofur with a stamp, a liquid trap is created. Droplets within this triangular confinement are not able to break through. **B** Destroying the nanohairs by scratching makes the sample hydrophilic again. These zones serve now as reservoirs for water. The undestroyed areas remain water repellent. **C** Since aqueous solutions can be manipulated easily, the water repellent nanofur may serve as substrate for experiments with liquids. As shown in this figure, 'Turnbull's Blue' was created by manipulating small amounts of the reagents without any residue.

For example, a microreactor can be easily fabricated using the nanofur as a substructure, where reservoirs and channels can be easily created by damaging the surface.

6.3.2. Slippery Liquid Infused Surfaces (SLIPS)

As demonstrated above, damaging the nanofur surfaces opens up a path to new applications in fluidics. However, there is one disadvantage apparent, which is the negative effect of wear on the superhydrophobicity. Addressing this challenge, Wong *et al.*[153] reported self-healing, slippery liquid infused surfaces (SLIPS) that repell various liquids like water or oil. Their approach was inspired by *Nepenthes* pitcher plants.[177] This insect-eating plant does not take advantage of surface structuring directly but rather uses microstructures in order to lock-in an intermediary liquid at its peristome. This intermediary liquid is slippery for insects. In this way, visiting insects slip and fall into the pitcher plant's trap.[177]

In case of SLIPS, the intermediary liquid is acting as water and oil repellent surface. The substrate and its surface roughness have to be adapted to ensure that its surface is preferentially wetted by the lubricating liquid rather than the liquid that has to be repelled.[153] Wong *et al.*[153] generated SLIPS by liquid imbibition of low-surface-tension perfluorinated liquids into porous materials. A random network of PTFE nanofibrous membranes or epoxy resin based nano-posts requiring the costly fabrication of a silicon master serve as substrate in this case.[178]

The presented hot pulling technique opens up the possibility to fabricate SLIPS very cost-effectively. In a proof-of-principle study, a polycarbonate sample densely covered with nanohairs was coated by putting it together with an object slide having 100µl of 1H,1H,2H,2H-perfluorodecyltrichlorosilane (FDTS, from Sigma-Aldrich Co. LLC.) on top into a vacuum desiccator for 15 hours. Fluorinert FC-70 (from 3M) was chosen as lubricating fluid . After applying droplets of FC-70 with a pipette to the nanofur, excessive FC-70 was shaken off. In order to visualize the existence of the lubricating fluid, n-Hexadecane was dispensed on the fabricated SLIPS and on the reference nanofur without the intermediary liquid (Figure 6.8A). The test liquid completely wet the nanofur whereas on the SLIPS a n-Hexadecane droplet was formed proving the existence of the lubricating fluid.

In case of SLIPS the wetting behavior is dominated by the lubricating fluid and not by the nanofur. The hairs of the substrate lock the lubricant on the surface.[153] Thus, physical damage of the surface topography by wear

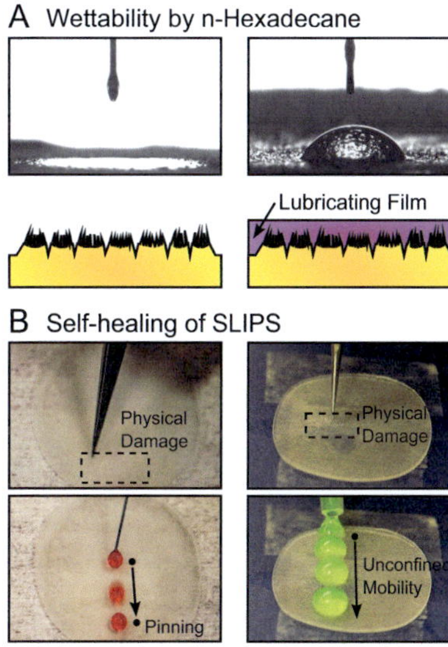

Figure 6.8.: A Wettability of a nanofur reference sample and a SLIPS surface by n-Hexadecane. On the reference sample the dispensed n-Hexadecane entirely spreads in milliseconds. On the SLIPS sample n-Hexadecane forms a droplet proving the existence of the intermediary liquid film consisting of a low-surface-tension perfluorinated liquid. **B** Composite photographs showing droplet motion after physical damage. After damaging the surface topography of the reference sample, a dispensed droplet pins at the defect. However, the lubricating liquid self-heals the physical damage on the SLIPS by refilling the damaged topography. Consequently, droplets roll off easily without pinning to the SLIPS.

is replenished by the lubricant resulting in the rapid restore of the liquid-repellent function. In Figure 6.8B the self-healing functionality of SLIPS is revealed. Physical damage of the surface topography does not negatively affect the droplet motion. This was because surface-energy-driven capillary action enforces the lubricant to refill the defect.[179] However, damaging the virgin nanofur without lubricating film results in an entirely modified characteristic in the defect zone. In this case, it destroys the hydrophilic property of polycarbonate resulting in the pinning of droplets in this area as shown before.

Consequently, hot pulling of nano-fur is a cost-effective method to fabricate a substrate that is able to lock-in fully-fluorinated liquids like Fluorinert FC-70. In this way, self-healing surfaces can be created offering wear-resistant superhydrophobicity.

6.3.3. Salvinia-Effect

Air-retaining surfaces are of great interest as they offer many applications, e.g., for low friction fluid transport and drag reducing ship coatings.[10,180–183] Several animals take advantage of an air-film that covers them underwater. The backswimming water bug *Notonecta glauca*, for instance, possesses a hierarchical formation of very small microtrichia with setae on top that covers their outer elytra.[184] This hierarchically structured surfaces holds a robust air-film that significantly reduces drag at the air-water interface. In this way, the water bug benefits while hunting, ensures dry wings, and gains air for respiration.[184] In flora, the leaves of the floating water fern *Salvinia* are the most popular example of surfaces that have the ability to retain an air layer when submerged in water.[10] Thus, this leaves have a silver shimmer underwater resulting from the reflection at the interface of water and trapped air (Figure 6.9A). Once emerged from water, the leaves are completely dry indicating their superhydrophobicity. Retaining an air film underwater enhances their buoyancy[11] and ensures sufficient respiration.[10] As CO_2 diffusion into water is much less compared to air[185] the advantage of retaining an air film enables the stomata of Salvinia to continue gas exchange.[11]

In order to maintain a robust air film from a few days (*Salvinia minima*)[11] to several weeks (*Salvinia molesta*)[10] an elaborate composition of microstructures covers the upper side of the floating leaves. Multicellular hairs branching into smaller hairs at their ends are covered with wax crystals, which make them hydrophobic. Depending on the species, these branches eventually join and form a flat patch at their terminal end (*Salvinia*

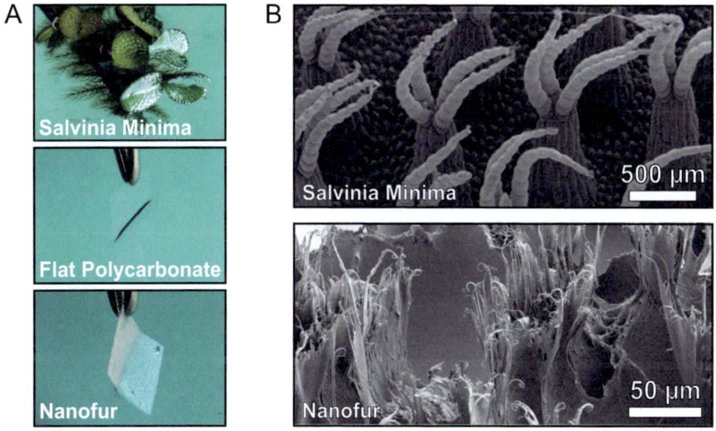

Figure 6.9.: **A** SEM images of the topography of Salvinia minima and the investigated nanofur. The leaves of *Salvinia Minima* exhibit quadruple trichomes, with length up to 800 µm. The leaf itself is around 5 mm in size. In early experiments performed by Cerman *et al.*[11] a submerged *Salvinia Minima* maintained a film of trapped air for 4-5 days. The topography of the chosen nanofur is hierarchical as well. The base of each micro-post is around 50 µm wide, the width of the nanohairs on top is less than 200 nm. **B** Photographs of *Salvinia minima* (reprinted with kind permissions from Barthlott *et al.*), a nanofur and a flat polycarbonate foil submerged into dyed water. The surface of the Salvinia leaf as well as the nanofur shimmer due to light reflection at the interface between water and trapped air. This distinctive effect is not observed for flat polycarbonate foils.

molesta).[10,11] In contrast to the others, these terminal cells are smooth and hydrophilic. During submersion, this unique combination ensures the stabilization of the air-water interface on top of the hairs.[10] Local currents and pressure fluctuations caused by turbulences fail to collapse the air-water interface. The hydrophobic wax coating prevents water from penetrating between the hairs, and at the same time, the hydrophilic cells are pinning the water, inhibiting removal of the interface.[10]

Exploiting air films underwater significantly reduces shear stress and, therefore, the surface friction drag, making appropriate surfaces interesting for commercial applications. However, technical air-retaining surfaces

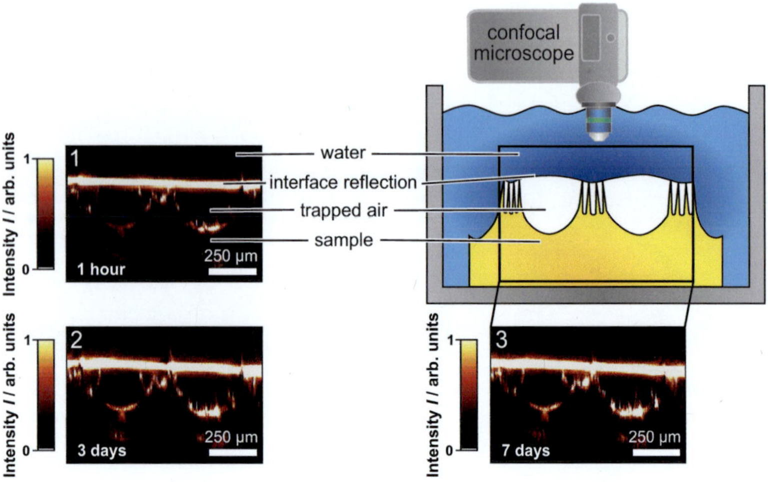

Figure 6.10.: The experimental setup and the vertical intensity profiles after 1 hour, 3 days and 7 days after submerging the nanofur into water are shown. In the intensity profiles, bright pixels indicate an interface. The lower interface separates air from the solid sample, whereas the upper bright zone shows the air-water interface. As it can be seen from the intensity profiles, the air film is stable for weeks. The air film was still intact when the experiment was stopped after 31 days (see Figure 6.11).

typically fail to maintain an air film for a sufficient lapse of time as well as withstanding turbulent flows.[186, 187]

The unique combination of hydrophobicity and hydrophilicity stabilizing the air-water film of a submerged Salvinia leaf was the inspiration to test the air-retaining capability of the nanofur. When submerged underwater, analogous to the Salvinia leaf, a bright shimmer was visible indicating trapped air (Figure 6.9A). In order to visualize the existence of a trapped air film on top of the surface, underwater experiments were performed with a confocal scanning optical microscope (CSOM) imaging the interfaces from above (Figure 6.10). In these experiments, a nanofur sample was put inside a glass beaker filled with water (head of water $\sim 4\,\text{mm}$). A CSOM spatially records the intensity of the reflection of the focused illumination spot.[188]

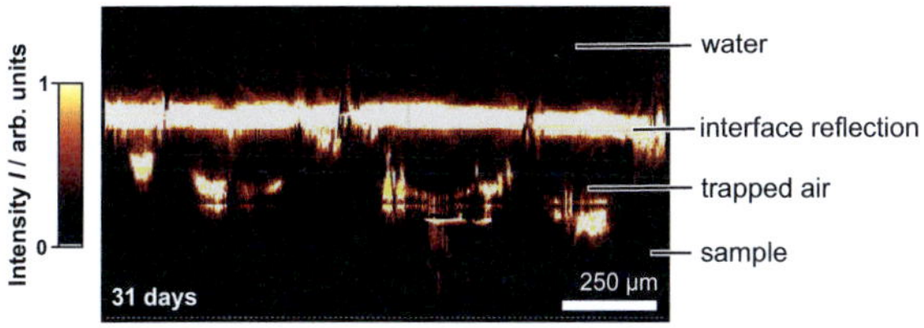

Figure 6.11.: Vertical intensity profile (CSOM) recorded 31 days after submerging the nanofur into water. The interfaces occuring bright in the raster image show the long-term stability of the retained air film.

Hence, it is very sensitive to interfaces as a difference in the refractive index of the mediums exists. Consequently, interfaces occur bright in the raster image. On the other hand, inside a medium diffusion dominates and less light is recorded - these fields occur dark in the raster image.

The vertical intensity profiles at 1 hour, 3 days and 7 days after submerging the nanofur are shown in Figure 6.10. Two different interfaces stand out. The lower interface retraces the topography of the nanofur. The upper interface separates water from trapped air. Single hairs of the nanofur break through this interface, serving as water pinning points and thus stabilizing the air-water interface. As can be seen from the intensity profiles, the interface is very stable in the long-term proving that the nanofur robustly retained air. Since the CSOM was not available for the complete 31 days, the intensity profile recorded after 31 days (Figure 6.11) shows a slightly different position than the measurements shown in Figure 6.10. The apparent interfaces perfectly show the robustness and long-term stability of the retained air-film. The experiment was halted after 31 days. Consequently, the nanofur is promising for underwater drag-reduction applications.

6.3.4. Oil/Water Separation

As reported by Ribeiro *et al.*[189] the hydrophobic leaves of Salvinia are applicable for oil/water filtration. As can be seen from Figure 6.8A the untreated nanofur is not only superhydrophobic but also superoleophilic. Combined with the high surface area these foils are ideal for the utilization as adsorbent for the oil/water separation.[190] In particular, the clean up and recovery of oil spills is a technical challenge that recently moved again in the focus of public attention in the context of the *Deepwater Horizon* blowout - the largest accidental marine oil spill in U.S. history.[191]

Throughout the explosion of the *Deepwater Horizon* drilling rig more than four million barrels of oil gushed uncontrolled into the Gulf of Mexico leading to a gigantic environmental disaster. Still, technology and practices for cleaning up oil spills lag behind the real risks of offshore deepwater drilling.[192] According to the *National Commission on the BP Deepwater Horizon Oil Spill and Offshore Drilling* this industrial accident has had enormous impact on the natural systems and economic losses of tens of billions of dollars.[192]

The collection of oil from the water surface is one of three options to clean up an oil spill. Alternative solutions are the enhancement of natural degradation by dispersants, and the in-situ burning of the oil spill.[193] However, both options preclude proper disposal of the oil.[193] In order to collect the oil from the water surface, absorbents are often applied to increase efficiency.[193] Usually, natural sorbents like sawdust are chosen to absorb the oil. However, the major drawbacks of these natural sorbents are the relatively low oil sorption capacity and low hydrophobicity leading to a large amount of coincidentally absorbed water.[190,193] Therefore, the absorbed oil/water mixture has to be reprocessed in order to separate oil from water. The applied methods like sedimentation, flotation, and centrifugation, however, encounter difficulties in separating emulsion particles with sizes smaller than 150 μm.[194] Adressing this challenge, elaborate solutions based on superoleophilic carbon nanotubes have to be applied.[194]

To avoid costly reprocessing, selective sorbents that prevent water from infiltrating are required. Mineral absorbents like zeolite and silica aerogels are amphiphilic (hydrophilic and oleophilic) at first but can be hydrophobized by appropriate treatment.[190,195–197] The high oil uptake capacity of these absorbents is outweighed by inflammability, brittleness, and non-permanent hydrophobicity.[190,193] The absorption of water may even cause silica aerogels to collapse.[190,195] Overcoming these drawbacks, synthetic organic sorbents such as commercial polypropylene or polyurethane fiber mats offer a high uptake capacity, too. However, when retracting these polymer absorbents the retention of oil is limited to about 50 %.[198] Hence, there is a substantial interest for sorbents providing high and rapid uptake combined with sufficient oil retention.

In order to investigate the applicability of the nanofur in oil/water separation an oil spill was simulated by dispensing oil (dyed with oil paint) in a beaker glass filled with water. To increase conspicuity, a thin layer of the undyed oil was dispensed first, so that the dyed oil spread less resulting in enhanced visibility. Subsequently, the nanofur was dipped into the contaminated water and was retracted afterwards. The experiment was repeated with different non-polar liquids (n-hexadecane and hydraulic oil 'Total Azolla ZS 10'). As shown in Figure 6.12, the oil was adsorbed by

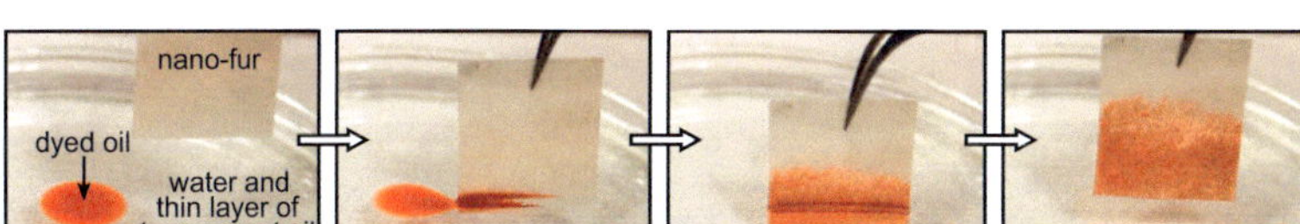

Figure 6.12.: Clean up of an oil spill by dipping the nanofur into the beaker glass filled with water. The oil is adsorbed whereas water is repelled enabling the oil/water separation.

the nanofur. The adsorbed colored oil climbed the nanofur beyond the water level. When retracting, the oil remained locked to the nanofur and was properly separated from the clean water. Thus, the non-polar oil is attracted by the nanofur whereas polar water is repelled. In this way, an oil uptake of up to $150\,\mathrm{mL\,m^{-2}}$ ('Total Azolla ZS 10') was measured.

The tendency of a liquid to totally wet a surface is described by the so-called *spreading parameter* S quantifying the difference of the surface energy (per unit area) of the dry and wet substrate[176]

$$S = \gamma_{SV} - (\gamma_{SL} + \gamma_{LV})\,. \tag{6.5}$$

In case of a positive spreading parameter ($S > 0$) the liquid spreads completely and forms a thin film on top of the substrate. If the spreading parameter is negative ($S < 0$) the liquid wets the surface only partially. Hence, with decreasing spreading parameter the wettability of a surface decreases. Contact angles of the resulting droplets below $90°$ ($\theta < 90°$) refer to wettable surfaces whereas contact angles above $90°$ ($\theta > 90°$) identify non-wettable surfaces. Using $\theta = 90°$ as boundary condition, Young's equation[155]

$$cos(\theta = 90°) = 0 = \frac{\gamma_{SV} - \gamma_{SL}}{\gamma_{LV}} \tag{6.6}$$

leads to $\gamma_{SL} = \gamma_{SV}$. Inserting this correlation into equation (6.5) results in the boundary spreading parameter S_b, giving the boundary for wettability and non-wettability

$$S_b = -\gamma_{LV}\,. \tag{6.7}$$

Interestingly, the boundary spreading parameter is independent of the substrate.

The surface tension of a liquid is composed of two contributing components: dispersive interactions (London dispersion forces) and polar interactions (i.e. hydrogen bonds or dipole-dipole interactions).[199] Polar molecules attract each other stronger than non-polar molecules. Therefore, the forces pulling the interfacial molecules inwards are higher for polar liquids, usually leading to a higher surface tension compared to non-polar liquids. With knowledge of the interfacial surface tensions of the substrate/vapor and liquid/vapor interface, according to Owens and Wendt[200] the corresponding interfacial surface tension of the substrate/liquid interface is

$$\gamma_{SL} = \gamma_{SV} + \gamma_{LV} - 2\sqrt{\gamma_{SV}^{d} \cdot \gamma_{LV}^{d}} - 2\sqrt{\gamma_{SV}^{p} \cdot \gamma_{LV}^{p}}, \tag{6.8}$$

where superscripts d and p represent the disperse and polar contribution, respectively. Therefore, knowledge of the substrates surface energy and the liquids surface tension is sufficient in order to calculate the spreading parameter S and to make a prediction of the wetting behavior. Defining the polar contribution ratio $p = \gamma_{LV}^{p}/\gamma_{LV}$ and combining eqs. (6.5) and (6.8) leads to

$$S = -2\gamma_{LV} + 2\sqrt{\gamma_{SV}^{d} \cdot (1-p)\,\gamma_{LV}} + 2\sqrt{\gamma_{SV}^{p} \cdot p \cdot \gamma_{LV}^{p}}. \tag{6.9}$$

The surface energy of the flat polycarbonate was determined to be $\gamma_{SV} = 46.78\,\mathrm{mN\,m^{-1}}$ and clearly dominated by dispersive interactions ($\gamma_{SV}^{d} = 46.76\,\mathrm{mN\,m^{-1}}$, $\gamma_{SV}^{p} = 0.02\,\mathrm{mN\,m^{-1}}$, OWRK method,[201] for details please see Appendix A.2).

In Figure 6.13 the spreading parameter for the flat polycarbonate is plotted for several surface tensions γ_{LV} as a function of their polar component. The chosen surface tensions correspond to the liquids listed in Table 6.2. As can be seen from the graph, the spreading parameter and therefore wettability decreases with increasing surface tension of the liquid. Interestingly the percentage of polarity contribution affects the spreading

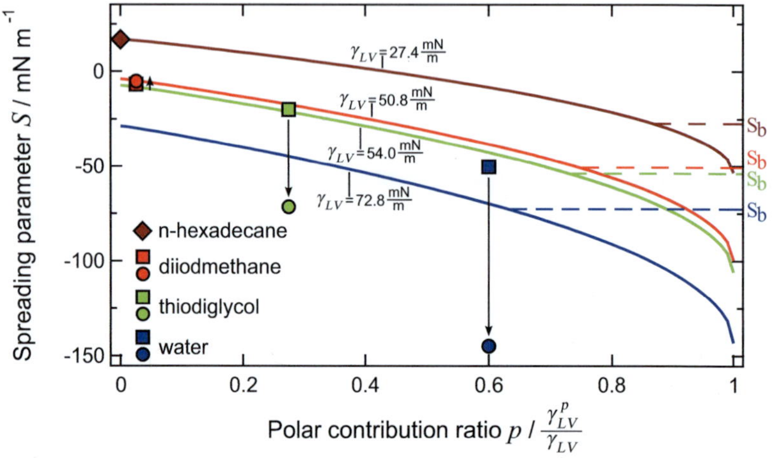

Figure 6.13.: Spreading parameter as a function of polar contribution to the surface tension. Theoretical spreading parameters (lines) are plotted for 4 different surface tensions. Data measured on a flat polycarbonate foil are indicated by squares (diiodmethane: red squar, thiodiglycol: green square, water: blue squares) and data measured on the nanofur are indicated by circles. The theoretical spreading parameter of n-hexadecane on a flat polycarbonate surface is marked with a brown diamond.

parameter even stronger. With increasing polar instead of dispersive interactions, the spreading parameter drops. Thus, the distribution of polar and disperse contributions to the surface tension have a very strong impact on the wettability of the flat polycarbonate.

The spreading parameters of diiodmethane, thiodiglycol and water were calculated by inserting the measured static contact angles into

$$S = \gamma_{LV} \left(cos\theta - 1 \right), \tag{6.10}$$

a combination of Young's equation[155] and equation (6.5). These calculated spreading parameters were added to the plot shown in Figure 6.13 (quadrangles with the colors: diiodmethane: red, thiodiglycol: green, water: blue). Interestingly, they match theory (lines) very well.

Furthermore, the theoretical spreading parameter of n-hexadecane was added to the graph (brown diamond). All these data points are well above their corresponding boundary spreading parameter S_b indicating that all tested liquids wet the flat polycarbonate very well.

For the nanofur serving as substrate, the spreading parameters of the test liquids were measured analogously (circles). Interestingly, the changes in the spreading parameter correspond to the polar contribution to the surface tension. Compared to the values measured for the flat polycarbonate the spreading parameter of the nearly non-polar diiodmethane slightly increased, whereas the spreading parameter decreased for the more polar liquids thiodiglycol and water. With increasing polar contribution, the change in the spreading parameter becomes larger. The spreading parameters of thiodiglycol and water are even below the corresponding boundary spreading parameters S_b and clearly in the non-wettable state.

Hence, the nearly non-polar liquids wet the nanofur even more whereas polar liquids are repelled. Consequently, the non-polar oil has a strong tendency to wet the nanofur whereas polar liquids like water are repelled enabling the oil/water separation by the nanofur.

This effect can also be applied for the filtration of emulsion consisting of a polar and a non-polar liquid. In the demonstration shown in Figure 6.14, an emulsion of inked water (blue) and dyed oil (orange) was applied onto a tilted nanofur and the flat polycarbonate reference. The emulsion dispensed

Table 6.2.: Surface tensions of test liquids and their dispersive and polar contributions.

liquid	γ_{LV} (mN m^{-1})	γ_{LV}^{d} (mN m^{-1})	γ_{LV}^{p} (mN m^{-1})	p
n-Hexadecane[202]	27.5	27.5	0.0	0.00
Diiodmethane[200]	50.8	49.5	1.3	0.03
Thiodiglycol[203]	54.0	39.2	14.8	0.27
Water[203]	72.8	29.1	43.7	0.60

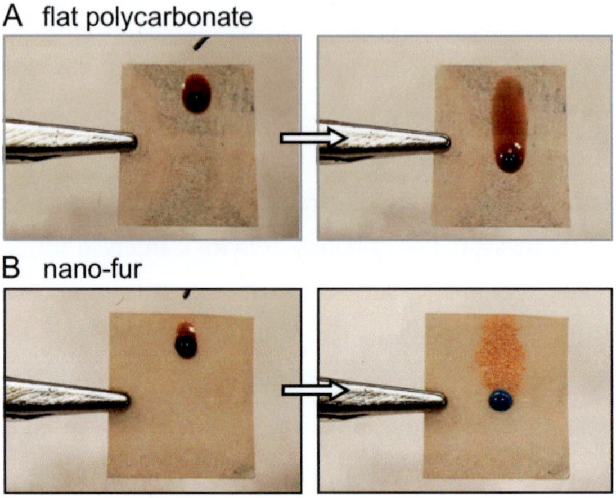

Figure 6.14.: Filtration of an oil/water emulsion. An emulsion of inked water (blue) and dyed oil (orange) is dispensed to a flat polycarbonate reference and the nanofur. In contrast to the flat foil, the nanofur separated oil from water by adsorbing and locking oil to the nano-hair.

to the flat polycarbonate didn't separate while running along the sample. The oil wets the flat surface and the rest of the emulsion slides on top of this film without segregation. In contrast, the emulsion applied to the nanofur segregates since the oil is adsorbed and locked by the nanohairs whereas the water rolls off. In this way, the water purification can be seen from the color change from black to blue.

6.4. Conclusion and Outlook

In summary, hot pulling was presented as a technique for the cost-effective fabrication of high aspect ratio nanohairs. This nanofur increases the static contact angle of polycarbonate from 72° up to 174°. By carefully choosing the hot pulling parameters droplets either slide or pin on the nanofur. Destroying the nanofur once again makes the surface hydrophilic in these areas opening a path to novel applications in microfluidics. Liquid traps and reservoirs can be easily created just by destroying the nanofur appropriately. In addition, high contact angles and low sliding angles make the nanofur perfect to serve as substrate for experiments with liquids, allowing residue free manipulation of droplets.

Destroying the nanofur by wear may affect the wetting of the surface undesirably. To overcome this problem, a SLIPS (slippery liquid infused surface) was created by locking a lubricating fluid within the nanofur. In this case, physical damage of the surface does not negatively affect hydrophobicity since the lubricant dominating the wetting behavior refills the defect.

Additionally, different wetting behavior for polar and non-polar liquids makes the nanofur applicable for the separation of oil/water emulsions. Simulating an oil spill, water was successfully purified by cleaning with the nanofur. The superhydrophobicity combined with superoleophilicity opens fields of application in filtration of polar and non-polar liquids.

Furthermore, the nanofur is capable of retaining air when submerged underwater. In the performed experiments it was shown that the retained air-film was highly robust for at least 31 days. This may pave the way to drag reduction in underwater applications.

The simplicity, scalability and the use of sandblasted steel plates as mold insert makes the presented hot pulling process applicable on the laboratory scale as well as for the insdustrial fabrication of superhydrophobic surfaces. Showing the variety of applications in an uncapped lab-on-a-chip device is one of the next steps to utilize the surfaces in practice.

A patent applications was submitted that refers to the fabrication and design of superhydrophobic surfaces, and the fabrication and design of smart surfaces used for oil/water separation. M. Röhrig, M. Schneider, M. Worgull and Hendrik Hölscher: Patent Application 102012111089.1 (2012).

A shortened version of this chapter was submitted as the article "Hot Pulling of Nanofur for Biomimetic Applications". M. Röhrig, M. Mail, M. Schneider, H. Louvin, A. Hopf, T. Schimmel, M. Worgull and Hendrik Hölscher (2013).

7. Conclusion and Outlook

Nature has taught us how micro- and nanostructures can be utilized to achieve exceptional effects. The surface effects described in this work such as the adhesion of gecko toes, the water repellency of lotus leaves, and the underwater air retaining ability of salvinia leaves are just a few examples of nature's patents. Innovative fabrication techniques allow manufacturing of ever smaller structure details. However, the fabrication of hierarchical high aspect ratio structures and the upscaling of these fabrication techniques are the major challenges that have to be mastered in order to get innovative bio-inspired products into the market.

In this work, novel micro- and nanostructuring techniques were introduced. The potentials of these techniques were exemplarily shown by means of gecko-inspired adhesives. Due to their hierarchical formation of high aspect ratio micro- and nanostructures the fabrication of such gecko-inspired adhesives is very challenging. By investigating the fabricated gecko-inspired adhesives, designs leading to highly sticky but self-cleaning adhesives were proposed. In this way, a gecko-like adhesive was presented that matches the adhesive strength and the self-cleaning ability of geckos very closely. Furthermore, the developed advanced molding processes were applied for the fabrication of superhydrophobic but superoleophilic nanofur that is water-repellent, air retaining and, when being surface treated, self-healing. The most relevant results of this work are briefly summarized below.

Geckos are easily able to climb on vertical walls or even ceilings. Due to the hierarchical design of their hairy attachment system, geckos achieve very intimate contact to flat as well as rough surfaces. In this way, the high contact area achieved with the substrate enables them to attach mainly with the help of van-der-Waals forces.

For the fabrication of synthetic gecko-like adhesives, however, it is not reasonable to copy geckos' attachment system 1:1. Rather, it is necessary to adopt the hierarchical micro- and nanostructures to the chosen material and fabrication process. Applying the iterative method discussed in Chapter 2, the design for robust gecko-like adhesives can be straightforwardly defined.

3D direct laser writing, a very flexible rapid prototyping technique, was applied to fabricate a design study of gecko-inspired micro- and nanostructures that vary in their order of magnitude, amount of hierarchical levels, pitch, aspect ratio and tip shape. In this way, the elastic modulus and dimensions of gecko setae were very closely matched. The adhesion measurements performed with an AFM and colloidal probes revealed that mushroom-shaped tips, high aspect ratios and hierarchy are indeed favorable for gecko-like adhesives.

Applying soft molding and dipping processes, different sizes of mushroom-shaped elastomeric microfibers were created. After contamination with different spherical glass particles, the contact self-cleaning ability was investigated by analyzing the recovered adhesion. It was observed that the self-cleaning ability depends on the size ratio of contaminants and microfibers. Microfibers much smaller than the contaminants allow the contaminants to roll across the adhesive during contact self-cleaning. Interestingly, recovering adhesion is not necessarily related to actually removing contaminants from the adhesive. Rather, embedding of the contaminants between the microfibers is the dominant cleaning mechanism that can be further enhanced by a hierarchical design. Exploiting these observations a gecko-like adhesive was presented that offers attachment strength (140 kPa) and adhesion recovery (up to 100 %) comparable to geckos.

The performed studies revealed that hierarchy, high aspect ratio and small scales are required to ensure high attachment strength and adequate self-cleaning ability of gecko-inspired adhesives. In order to present a scalable replication process, advanced hot embossing techniques were developed in this work. Using the introduced electromechanical sensor, hierarchical micro- and nanostructures can be precisely fabricated in a fully automated process. Applying the novel hot pulling technique allows for the fabrication of filigree high aspect ratio micro- and nanostructures. In this way, an aspect ratio of 10 was achieved for a regular array of nanostructures which are 140 nm in diameter. Combining these advanced molding techniques, threefold hierarchical micro and nanostructures that offer remarkable adhesion were fabricated (Figure 7.1).

Using sandblasted steel plates as mold inserts, hot pulling is a very cost-effective method for the fabrication of nanofur that is beneficial for various biomimetic applications. This easy and scalable fabrication method enables tuning the wettability of polycarbonate from hydrophilic to superhydrophobic. Changing the fabrication parameters results in superhydrophobic samples with either high or low adhesion to water. By damaging the surface structure, liquid traps were created by changing the local wettability. To overcome wearing issues, self-healing is achievable by locking an intermediary liquid acting as water and oil repellent surface.

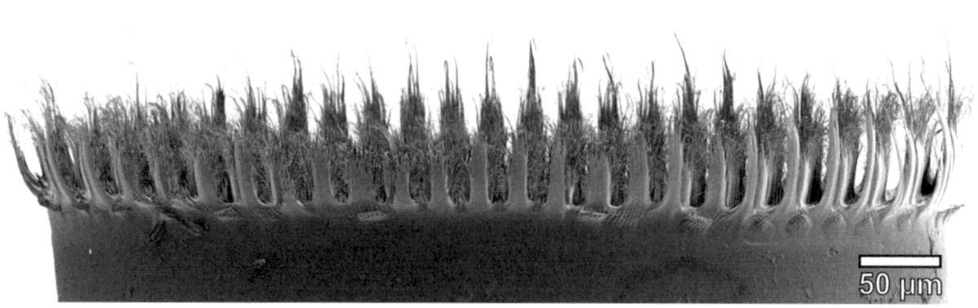

Figure 7.1.: Inspired by geckos, threefold hierarchical micro- and nanostructures were successfully fabricated by combining classical hot embossing, hierarchical hot embossing and hot pulling.

Mimicking the leaves of the floating water fern Salvinia the nanofur retains air when submerged underwater. Therefore, the nanofur is promising for drag reduction in underwater applications. Due to its superhydrophobicity and superoleophilicity, the nanofur is applicable for oil/water separation and oil spill clean up.

To successfully perform the transition from laboratory to the market, bioinspired smart surfaces have to be produced very cost-effectively. For this, the fabrication process has to be transferred to a roll-to-roll process as established for the macroscopic structuring of, e.g., reflective foils and Fresnel lenses.[204] As shown by numerous groups[166,167] upscaling of micro hot embossing processes is feasible by transferring them to roll-to-roll embossing, however, the curvature of the roll limits the achievable aspect ratio in classical roll-to-roll embossing. Another challenge is the attachment of microstructured mold inserts to the curvature of the roll. Possibly, differences in thermal expansion can be utilized to strongly attach the mold inserts to the underlying roll by heat shrinking. The task of attaching the mold insert can be avoided by directly micro- and nanostructuring the roll. For this, micromachining and etching processes are available.[205]

The compliance of the thermoplastic gecko-like adhesives can be enhanced by mimicking the gradient in elastic moduli as observed in the hairy attachment systems of beetles.[206] Soft endings could be added to the synthetic adhesives by using the dipping process presented in this work. In this way, the high adaptability and adhesion of low surface energy materials such as polyurethane or polydimethylsiloxane can be utilized to achieve a strong bond, whereas thermoplastic hierarchy below ensures compliance to milli- and microscale roughnesses and prevents structural collapse. Using the particular material class of thermoplastic elastomers that offer elastic moduli down to the megapascals range creates new opportunities to enhance compliance of gecko-like adhesive fabricated by advanced hot embossing processes. Possibly, by using an adequate material combination a stack of a thermoplastic polymer, a thermoplastic elastomer, and an

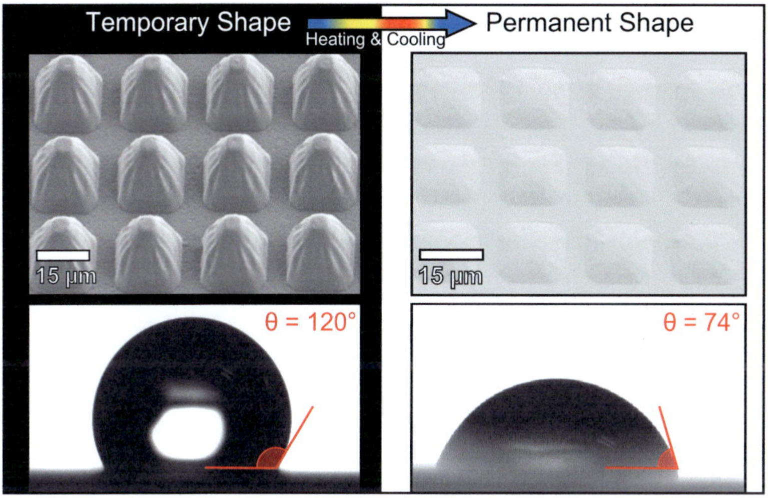

Figure 7.2.: Wetting transformation by switching a microstructured surface on demand. The temporary shape of the microstructured surface (left) shows a hydrophobic contact angle of (120°) caused by pyramidal pillars with a height of 25 µm and an edge length of 17 µm. After heating, the surface recovers its permanent flat shape (right) and the microstructure disappears. Consequently, the contact angle of the cooled surface reduces to that of a flat surface (72°).

elastomer lying upon each other can be used as the semifinished sheet.

In general, novel material classes such as shape memory polymers or liquid wood will extend the functionality and environmental compatibility of bio-inspired products. Shape memory polymers such as Tecoflex EG 72-D (Lubrizol Corporation) consist of a low and a high temperature phase. Hot embossing above the glass transition temperature of the high temperature phase programs the permanent shape. The temporary shape is defined by molding below the glass transition temperature but above the softening temperature of the low temperature phase and subsequent demolding below the softening temperature. As shown in Figure 7.2, microstructured shape memory polymers were morphed between their temporary and permanent shape triggered by heating to above the softening

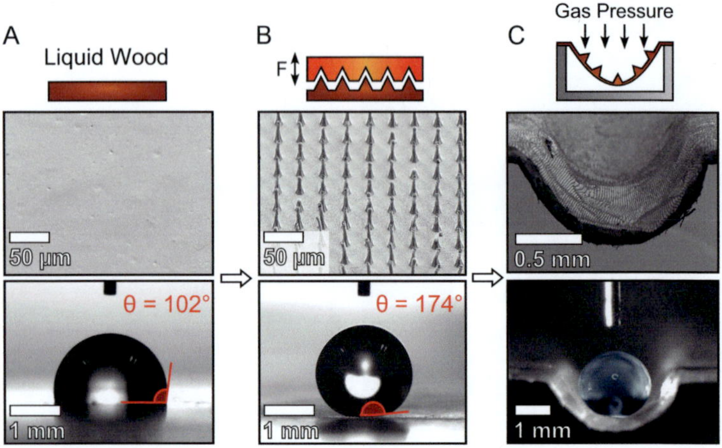

Figure 7.3.: Manufacturing of a superhydrophobic microfluidic channel from liquid wood. **A** A flat unstructured foil of liquid wood is hydrophobic with a contact angle of 102°. **B** By hot embossing above the softening temperature the surface can be microstructured. Due to the microstructure the surface revealed superhydrophobic properties with a contact angle of 174°. **C** By heating the microstructured foil near to the softening temperature, microthermoforming allows the fabrication of 3D shapes. Since the microthermoforming temperature is well below the molding temperature during hot embossing the surface structure is preserved. Consequently, the final microfluidic channel is still super-hydrophobic.

temperature. In this way, wetting can be controlled as shown in the lower part of Figure 7.2. In the temporary state, pyramidal pillars with a height of 25 µm and an edge length of 17 µm result in a hydrophobic contact angle of 120°. When heated above the softening temperature the surface morphs and recovers its flat permanent shape that is hydrophilic ($\theta = 72°$). Applying such shape memory polymers for the fabrication of gecko-inspired micro- and nanostructures leads to switchable adhesives.[116] In this way, fibers can collapse on demand to support easy detachment. This is a feature required for the cost-effective recycling of bonded components.

Recycling is a big topic in general. Per year more then 6.4 million tonnes of plastic waste end up in oceans.[207] Replacing the non-biodegradable polymers by environmentally compatible materials would drastically reduce this pollution. Liquid wood is a novel material glass that will prospectively loom large in this sector. This biopolymer is made of lignin making it fully compostable. At the Institute of Microstructuretechnology (IMT) liquid wood was successfully micro- and nanostructured for the first time applying hot embossing and thermoforming processes (Figure 7.3). In this way, superhydrophobic microchannels were fabricated out of pure wood.

Research and development of advanced micro- and nanofabrication processes and novel materials made enormous progress in the past 50 years. This trend is presumably set to continue bringing us a step closer to the marvelous biological systems, as already envisioned by Richard P. Feynman. Many more patents of nature will be discovered and corresponding innovations will successfully assist us in a wide range of fields.

A. Appendix

A.1. Experimental details of the fabrication of nano-fur (Chapter 6)

In order to fabricate the nano-fur, the polycarbonate Makrolon LED2045 (Bayer MaterialScience AG, Leverkusen, Germany) was used as polymer. The semi-finished polymer foils (thickness: 1 mm) were produced by hot embossing the dried polymer granulate (machine: Wickert WMP1000, embossing temperature: 170 °C, embossing force: 350 kN, demolding temperature: 120 °C). The polycarbonate foils were cleaned with isopropyl alcohol afterwards.

Hot pulling was performed with the hot embossing machine HEX03 (Jenoptik AG, Jena, Germany). Steel-plates (stainless steel X5CrNi18-10) were used as molds. In order to pattern the steel-plates, they were sand-blasted twice (1. aluminum silicate clinker 0.6 - 1.4 mm, 2. aluminum oxide 15.5 - 17.5 µm).

A.2. Experimental details of the surface energy and contact angle measurement (Chapter 6)

The measurements of surface energies and static contact angles were performed with the contact angle measurement system OCA 40 (DataPhysics Instruments GmbH, Filderstadt, Germany). To measure the static contact angles, water droplets with a volume of 4.5 µl were dispensed to the surfaces (DI-water with $\gamma_{LV} = 72.9 \pm 0.7\,\mathrm{mN\,m^{-1}}$ measured by pendant drop method; temperature $\approx$ 22 °C, rel. humidity $\approx$ 50 %, clean room conditions). Static contact angles were measured with the software SCA20 (DataPhysics Instruments GmbH, Filderstadt, Germany) applying Young-Laplace fitting. Advancing and receding contact angles were obtained from the sliding angle measurement data. Just before the droplets started to slide, the contact angles at front and rear were manually measured with the 'Drop Shape Analysis plugin' for the software 'ImageJ' (http://rsbweb.nih.gov/ij/). The surface energy of the flat polycarbonate was calculated applying the method according to Owens, Wendt, Rabel and Kaelble[201] (test liquids: ethyleneglycol, thiodiglycol and diiodmethane; droplet volume: 1 µL, temperature $\approx$ 24.5 °C, rel. humidity $\approx$ 15.5 %, clean room conditions).

B. List of Publications

B.1. Patents

1. M. Röhrig, M. Schneider, G. Etienne, M. Worgull, H. Hölscher. *Verfahren zur Herstellung eines Formkörpers mit mindestens zwei hierarchisch zueinander angeordneten Strukturebenen (Fabrication process for hierarchical structures that have not less than two hierarchical levels, and the design of these surfaces)*, PATENT APPLICATION **102012110048.9** (2012)

2. M. Röhrig, M. Schneider, M. Worgull, H. Hölscher. *Verfahren zur Herstellung eines Formkörpers mit superhydrophober Oberfläche und Formkörper erhältlich mit diesem Verfahren (Fabrication process for superhydrophobic surfaces, and the design of these surfaces)*, PATENT APPLICATION **102012111089.1** (2012)

B.2. Articles and Book Contributions

1. M. Röhrig, M. Heilig, M. Schneider, A. Kolew, K. Kaiser, M. Guttmann, and M. Worgull. *Universelle Fixierung von Shim-Formeinsätzen für die Mikro- und Nanoreplikation auf Basis von Polymerfolien*, GALVANOTECHNIK **101**, 1646 (2010)

2. M. Röhrig, M. Thiel, M. Worgull, and H. Hölscher. *3D Direct Laser Writing of Nano- and Microstructured Hierarchical Gecko-Mimicking Surfaces*, SMALL **8**, 3009 (2012)

3. L. Xiao, S.E.M. Thompson, M. Röhrig, M.E. Callow, J.A. Callow, M. Grunze, and A. Rosenhahn. *Hot Embossed Microtopographic Gradients Reveal Morphological Cues That Guide the Settlement of Zoospores*, LANGMUIR **29**, 1093 (2013)

4. M. Röhrig, M. Thiel, S. Bundschuh, M. Worgull, and H. Hölscher. *How Geometry Affects the Adhesion of Gecko-Like Adhesives*, Book Chapter in BIOLOGICAL AND BIOMIMETIC ADHESIVES: CHALLENGES AND OPPORTUNITIES ed. by S. Gorb, P. Flammang, R. Santos, N. Aldred, Royal Society of Chemistry Publishing (2013)

5. Y. Mengüç and M. Röhrig (equally contributing), U. Abusomwan, H. Hölscher, and M. Sitti. *Staying Sticky: Contact Self-Cleaning of Gecko-Inspired Adhesives*, submitted (2013)

6. M. Worgull, M. Schneider, M. Röhrig, T. Meier, M. Heilig, A. Kolew, K. Feit, H. Hölscher, and J. Leuthold. *Hot embossing and thermoforming of biodegradable three-dimensional wood structures*, RSC ADV. **3**, 20060 (2013)

7. M. Worgull, M. Reinhard, M. Röhrig, M. Schneider, T. Meier, M. Heilig, J. Leuthold, and H. Hölscher. *Morphing of Nano- and Microstructures: Switching of Functional Surfaces Using Shape Memory Polymers*, to be submitted

8. M. Röhrig, M. Schneider, G. Etienne, F. Oulhadj, F. Pfannes, M. Worgull and H. Hölscher. *Hot pulling and embossing of hierarchical nano-and micro-structures*, J. MICROMECH. MICROENG. **23**, 105014 (2013)

9. M. Röhrig, M. Mail, M. Schneider, H. Louvin, A. Hopf, T. Schimmel, M. Worgull, and H. Hölscher. *Hot Pulling of Nanofur for Biomimetic Applications*, submitted (2013)

B.3. Conference Contributions (Oral)

1. M. Röhrig, M. Heilig, A. Kolew, M. Schneider, H. Hölscher, and M. Worgull. *Hot Embossing for Biomimetics - Replication of Surface Structures in the Sub-Micron Range*, 1st SCIENTIFIC MEETING ON BIOLOGICAL AND BIOMIMETIC ADHESIVES, Vienna, Austria, November 04 - 05 (2010)

2. M. Röhrig, A. Kolew, F. Pfannes, M. Worgull, and H. Hölscher. *Artificial Hierarchical Gecko-mimicking Structures*, 75. JAHRESTAGUNG DER DPG, Dresden, Germany, March 13 - 18 (2011)

3. M. Röhrig, A. Kolew, F. Pfannes, M. Worgull, and H. Hölscher. *Fabrication of Hierarchical Gecko-mimicking Structures*, 2nd SCIENTIFIC MEETING ON BIOLOGICAL AND BIOMIMETIC ADHESIVES, Mons, Belgium, May 18 - 20 (2011)

4. M. Röhrig, F. Oulhadj, M. Thiel, M. Worgull, and H. Hölscher. *Hierarchical Structuring of Dry Adhesives by 3D Laser Lithography and Hot Pulling*, **invited**, GORDON-KENAN RESEARCH SEMINAR ON SCIENCE OF ADHESION, Lewiston, USA, June 23 - 24 (2011)

5. L. Xiao, M. Röhrig, S.E. Thompson, M.E. Callow, J.A. Callow, A. Rosenhahn, and M. Grunze. *Surface Topographic Features to control Biofouling*, AVS 58th INTERNATIONAL SYMPOSIUM & EXHIBITION, Nashville, USA, October 30 - November 04 (2011)

6. M. Röhrig, M. Thiel, M. Worgull, and H. Hölscher. *Rapid Prototyping of Hierarchical Gecko-Mimicking Nano- and Microstructures Using Multiphoton Absorption*, 76. JAHRESTAGUNG DER DPG, Berlin, Germany, March 25 - 30 (2012)

7. M. Röhrig, M. Thiel, M. Worgull, and H. Hölscher. *How Geometry Affects Dry Adhesion - A Systematic Design Study using 3D Direct Laserwriting*, 1th INTERNATIONAL CONFERENCE ON BIOLOGICAL AND BIOMIMETIC ADHESIVES, Lisbon, Portugal, May 09 - 11 (2012)

8. M. Röhrig, M. Thiel, M. Worgull, and H. Hölscher. *3D Direct Laser Writing of Micro- and Nanostructured Gecko-Like Adhesives*, MRS SPRING MEETING & EXHIBIT, San Francisco, USA, April 01 - 05 (2013)

9. M. Röhrig, Y. Mengüç, U. Abusomwan, H. Hölscher and M. Sitti. *Contact Self-Cleaning of Gecko-Inspired Elastomer Micro-Fibrillar Adhesives*, MRS SPRING MEETING & EXHIBIT, San Francisco, USA, April 01 - 05 (2013)

B.4. Conference Contributions (Poster)

1. M. Röhrig, M. Worgull, and H. Hölscher. *Hot Embossing in Biomimetics*, BIOINSPIRED ADHESION - FROM GECKOS TO NEW PRODUCTS, Saarbrücken, Germany, July 07 - 09 (2010)

2. M. Röhrig, A. Kolew, F. Pfannes, M. Worgull, and H. Hölscher. *AFM Analysis of Gecko-mimicking Polymer Structures*, 2nd INTERNATIONAL WORKSHOP ON ADVANCED ATOMIC FORCE MICROSCOPY, Karlsruhe, Germany, February 28 - March 01 (2011)

3. M. Röhrig, F. Oulhadj, M. Thiel, M. Worgull, and H. Hölscher. *Hierarchical Structuring of Dry Adhesives by 3D Laser Lithography and Hot Pulling*, GORDON RESEARCH CONFERENCE SCIENCE OF ADHESION, Lewiston, USA, June 24 - 29 (2011)

4. M. Röhrig, F. Oulhadj, M. Thiel, M. Worgull, and H. Hölscher. *Hierarchical Structuring of Dry Adhesives by 3D Laser Lithography and Hot Pulling*, **1st Prize Poster Award**, STATUSWORKSHOP DES KOMPETENZNETZES FUNKTIONELLE NANOSTRUKTUREN, Bad Herrenalb, Germany, September 14 - 16 (2011)

5. M. Röhrig, M. Thiel, F. Oulhadj, F. Pfannes, M. Worgull, and H. Hölscher. *Adhesion Analysis of Gecko-Inspired Hierarchical Adhesives Using Atomic Force Microscopy*, 14th INTERNATIONAL CONFERENCE ON NON-CONTACT ATOMIC FORCE MICROSCOPY, Lindau at Lake Constance, Germany, September 18 - 22 (2011)

6. L. Xiao, S. E. Thompson, M. Röhrig, M. E. Callow, J. A. Callow, A. Rosenhahn, M. Grunze. *Ulva Settlement on Surface Topographic Gradient*, INTERNATIONAL WORKSHOP OF THE US OFFICE OF NAVAL RESEARCH (ONR) AND SEACOAT, Las Vegas, USA, December 04 - 08 (2011)

7. M. Röhrig, M. Thiel, M. Worgull, and H. Hölscher. *Analysis of Gecko-Mimicking Hierarchical Nano- and Microstructures fabricated by direct laser writing*, 3rd INTERNATIONAL WORKSHOP ON ADVANCED ATOMIC FORCE MICROSCOPY, Karlsruhe, Germany, March 5 - 6 (2012)

8. G. Etienne, M. Röhrig, F. Oulhadj, M.Schneider, M. Worgull, and H. Hölscher. *Hierarchical Fabrication of Gecko-Mimicking Nano- and Microstructures by Hot Embossing and Hot Pulling*, 76. JAHRESTAGUNG DER DPG, Berlin, Germany, March 25 - 30 (2012)

9. M. Röhrig, G. Etienne, F. Oulhadj, M.Schneider, M. Worgull, and H. Hölscher. *Cost-Effective Fabrication of Hierarchical Gecko-Mimicking Structures*, 1th INTERNATIONAL CONFERENCE ON BIOLOGICAL AND BIOMIMETIC ADHESIVES, Lisbon, Portugal, May 09 - 11 (2012)

10. G.K. Darbha, J. Lützenkirchen, M. Röhrig, C. Fischer, T. Schäfer. *Understanding the Interaction Forces between Colloids and Surfaces as a Function of Surface Roughness and Metal Ion Concentration*, 4th INTERNATIONAL WORKSHOP ON ADVANCED ATOMIC FORCE MICROSCOPY, Karlsruhe, Germany, March 04 - 05 (2013)

11. J. Syurik, M. Röhrig, O.I. Ilin, A.A. Fedotov, O.A. Ageev, and H. Hölscher. *YCNTs- Based Gecko-inspired Adhesion Surfaces*, 4th INTERNATIONAL WORKSHOP ON ADVANCED ATOMIC FORCE MICROSCOPY, Karlsruhe, Germany, March 04 - 05 (2013)

B.5. Seminar Talks at other Institutes

1. M. Röhrig, F. Oulhadj, M. Thiel, M. Worgull, and H. Hölscher. *Hierarchical Structuring of Gecko-Inspired Dry Adhesives Fabricated by 3D Laser Lithography and Hot Pulling*, MECHANICAL ENGINEERING, Carnegie Mellon University, Pittsburgh, USA, October 10 (2011)

2. M. Röhrig, F. Oulhadj, M. Thiel, M. Worgull, and H. Hölscher. *Hierarchical Structuring of Gecko-Inspired Dry Adhesives Fabricated by 3D Laser Lithography and Hot Pulling*, MECHANICAL ENGINEERING & SCIENCE DEPT., Yale University, New Haven, USA, December 05 (2011)

List of Figures

List of Tables

Bibliography

[1] R. P. Feynman. There's plenty of room at the bottom - an invitation to enter a new field of physics. *Caltech Engineering and Science*, 23:5:22–36, 1960.

[2] M. Fuechsle, J. A. Miwa, S. Mahapatra, H. Ryu, S. Lee, O. Warschkow, L. C. L. Hollenberg, G. Klimeck, and M. Y. Simmons. A single-atom transistor. *Nature Nanotechnology*, 7(4):242–246, 2012.

[3] V. Stadler, T. Felgenhauer, M. Beyer, S. Fernandez, K. Leibe, S. Güttler, M. Gröning, K. König, G. Torralba, M. Hausmann, V. Lindenstruth, A. Nesterov, I. Block, R. Pipkorn, A. Poustka, F. R. Bischoff, and F. Breitling. Combinatorial synthesis of peptide arrays with a laser printer. *Angewandte Chemie International Edition*, 47(37):7132–7135, 2008.

[4] W. Nachtigall. *Bionik - Grundlagen und Beispiele für Ingenieure und Naturwissenschaftler*. Springer-Verlag Berlin Heidelberg, 2002.

[5] Z. Cerman, W. Barthlott, and J. Nieder. *Erfindungen der Natur: Bionik - Was wir von Pflanzen und Tieren lernen können*. Rowohlt Taschenbuch Verlag, 2011.

[6] J. M. Benyus. *Biomimicry: Innovation Inspired by Nature*. William Morrow Paperbacks, 2002.

[7] Bundesministerium für Bildung und Forschung (BMBF). Innovationen aus der Natur - Förderkonzept Bionik, 2005.

[8] C. Neinhuis and W. Barthlott. Characterization and distribution of water-repellent, self-cleaning plant surfaces. *Annals of Botany*, 79(6):667–677, 1997.

[9] W. Barthlott and C. Neinhuis. Purity of the sacred lotus, or escape from contamination in biological surfaces. *Planta*, 202(1):1–8, 1997.

[10] W. Barthlott, T. Schimmel, S. Wiersch, K. Koch, M. Brede, M. Barczewski, S. Walheim, A. Weis, A. Kaltenmaier, A. Leder, and H. F. Bohn. The salvinia paradox: Superhydrophobic surfaces with hydrophilic pins for air retention under water. *Advanced Materials*, 22(21):2325–2328, 2010.

[11] Z. Cerman, B. F. Striffler, W. Barthlott, and edited by S. N. Gorb. *Functional Surfaces in Biology - Little Structures with Big Effects, Dry in the water: The superhydrophobic water fern Salvinia - A model for biomimetic surfaces*. Springer Science + Business Media, 2009.

[12] W. R. Hansen and K. Autumn. Evidence for self-cleaning in gecko setae. *PNAS*, 102:385, 2005.

[13] S. Hu, S. Lopez, P. H. Niewiarowski, and Z. Xia. Dynamic self-cleaning in gecko setae via digital hyperextension. *Journal of The Royal Society Interface*, 2012.

[14] D. J. Irschick, C. C. Austin, K. Petren, R. N. Fisher, J. B. Losos, and O. Ellers. A comparative analysis of clinging ability among pad-bearing lizards. *Biological Journal of the Linnean Society*, 59:21–35, 1996.

[15] K. Autumn, M. Sitti, Y. A. Liang, A. M. Peattie, W. R. Hansen, S. Sponberg, T. W. Kenny, R. Fearing, J. N. Israelachvili, and R. J. Full. Evidence for van der waals adhesion in gecko setae. *PNAS*, 99(19):12252–12256, 2002.

[16] Copyright Eric Isselée, 2012. Used under license from Shutterstock.com.

[17] Kellar Autumn. How gecko toes stick. *American Scientist*, March-April 2006:123–132, 2006.

[18] D. Quéré. Rough ideas on wetting. *Physica A: Statistical Mechanics and its Applications*, 313:32 – 46, 2002.

[19] W.-D. Dellit. Zur Anatomie und Physiologie der Geckozehe. *Jenaische Zeitschrift für Naturwissenschaft*, 68:613–656, 1934.

[20] R. Ruibal and V. Ernst. The structure of the digital setae of lizards. *Journal of Morphology*, 117(3):271–293, 1965.

[21] U. Hiller. Untersuchungen zum Feinbau und zur Funktion der Haftborsten von Reptilien. *Zeitschrift für Morphologie der Tiere*, 62:307–362, 1968.

[22] K. Autumn, Y. A. Liang, S. T. Hsieh, W. Zesch, W. P. Chan, T. W. Kenny, R. Fearing, and Robert J. Full. Adhesive force of a single gecko foot-hair. *Nature*, 405(6787):681–685, 2000.

[23] A. Nakajima, K. Hashimoto, and T. Watanabe. Recent studies on super-hydrophobic films. *Monatshefte für Chemie*, 132:31–41, 2001.

[24] A. Asbeck, S. Dastoor, A. Parness, L. Fullerton, N. Esparza, D. Soto, B. Heyneman, and M. Cutkosky. Climbing rough vertical surfaces with hierarchical directional adhesion. In *Robotics and Automation, 2009. ICRA '09. IEEE*, pages 2675–2680, 2009.

[25] M. P. Murphy, C. Kute, Y. Menguc, and M. Sitti. Waalbot II: adhesion recovery and improved performance of a climbing robot using fibrillar adhesives. *The International Journal of Robotics Research*, 30(1):118–133, 2010.

[26] H. E. Jeong, J.-K. Lee, H. N. Kim, S. H. Moon, and K. Y. Suh. A nontransferring dry adhesive with hierarchical polymer nanohairs. *PNAS*, 106(14):5639–44, 2009.

[27] Y. Menguc, S. Y. Yang, S. Kim, J. A. Rogers, and M. Sitti. Gecko-inspired controllable adhesive structures applied to micromanipulation. *Advanced Functional Materials*, 2011.

[28] J. Eilperin. Innovators develop products that mimic nature's principles. *Washington Post*, December 29, 2008.

[29] Fermanian Business & Economic Institute. *The Global Biomimicry Efforts: An Economic Game Changer*. Comissioned by San Diego Zoo Global, 2010.

[30] L. Xiao, S. E. M. Thompson, M. Röhrig, M. E. Callow, J. A. Callow, M. Grunze, and A. Rosenhahn. Hot embossed microtopographic gradients reveal morphological cues that guide the settlement of zoospores. *Langmuir*, 29(4):1093–1099, 2012.

[31] K. Koch, B. Bhushan, Y. C. Jung, and W. Barthlott. Fabrication of artificial lotus leaves and significance of hierarchical structure for superhydrophobicity and low adhesion. *Soft Matter*, 5:1386–1393, 2009.

[32] B. Aksak, M. P. Murphy, and M. Sitti. Adhesion of biologically inspired vertical and angled polymer microfiber arrays. *Langmuir*, 23(6):3322–3332, 2007.

[33] A. del Campo, C. Greiner, I. Álvarez, and E. Arzt. Patterned surfaces with pillars with controlled 3d tip geometry mimicking bioattachment devices. *Advanced Materials*, 19(15):1973–1977, 2007.

[34] L. Ge, S. Sethi, L. Ci, P. M. Ajayan, and A. Dhinojwala. Carbon nanotube-based synthetic gecko tapes. *PNAS*, 104(26):10792–10795, 2007.

[35] S. Gorb, M. Varenberg, A. Peressadko, and J. Tuma. Biomimetic mushroom-shaped fibrillar adhesive microstructure. *Journal of The Royal Society Interface*, 4(13):271–275, 2007.

[36] C. Greiner, E. Arzt, and A. del Campo. Hierarchical gecko-like adhesives. *Advanced Materials*, 21(4):479–482, 2009.

[37] G. Pötsch and W. Michaeli. *Injection Molding*. Hanser Gardner Publications, 1995.

[38] Michael D. Austin, Haixiong Ge, Wei Wu, Mingtao Li, Zhaoning Yu, D. Wasserman, S. A. Lyon, and Stephen Y. Chou. Fabrication of 5 nm linewidth and 14 nm pitch features by nanoimprint lithography. *Applied Physics Letters*, 84(26):5299–5301, 2004.

[39] M. Worgull. *Hot Embossing - Theory and Technology of Microreplication*. William Andrew Applied Science Publishers, 2009.

[40] M. Heckele and W. K. Schomburg. Review on micro molding of thermoplastic polymers. *Journal of Micromechanics and Microengineering*, 14(3):R1, 2004.

[41] Alexander Kolew, Daniel Münch, Karsten Sikora, and Matthias Worgull. Hot embossing of micro and sub-micro structured inserts for polymer replication. *Microsystem Technologies*, 17:609–618, 2011.

[42] T. Mappes, M. Worgull, M. Heckele, and J. Mohr. Submicron polymer structures with X-ray lithography and hot embossing. *Microsystem Technologies*, 14:1721–1725, 2008.

[43] M. P. Murphy, S. Kim, and M. Sitti. Enhanced adhesion by gecko-inspired hierarchical fibrillar adhesives. *ACS Applied Materials & Interfaces*, 1(4):849–855, 2009.

[44] P. Y. Hsu, L. Ge, X. Li, A. Y. Stark, C. Wesdemiotis, P. H. Niewiarowski, and A. Dhinojwala. Direct evidence of phospholipids in gecko footprints and spatulae substrate contact interface detected using surface-sensitive spectroscopy. *Journal of The Royal Society Interface*, 9(69):657–664, 2012.

[45] K. Autumn, C. Majidi, R. E. Groff, A. Dittmore, and R. Fearing. Effective elastic modulus of isolated gecko setal arrays. *The Journal of Experimental Biology*, 209(18):3558–3568, 2006.

[46] K. Autumn and A. M. Peattie. Mechanisms of adhesion in geckos. *Integrative and Comparative Biology*, 42(6):1081–1090, 2002.

[47] N. S. Pesika, H. Zeng, K. Kristiansen, B. Zhao, Y. Tian, K. Autumn, and J. Israelachvili. Gecko adhesion pad: a smart surface? *Journal of Physics: Condensed Matter*, 21(46):464132, 2009.

[48] J. Batal. Saarbrücker Forscher zeigen Gecko-Effekt an künstlicher Oberfläche. *campus*, 01/2006:21, 2006.

[49] S. N. Gorb. *Attachment Devices of Insect Cuticle*. Springer, 2001.

[50] S. N. Gorb, M. Sinha, A. Peressadko, K. A. Daltorio, and R. D. Quinn. Insects did it first: a micropatterned adhesive tape for robotic applications. *Bioinspiration & Biomimetics*, 2(4):S117–S125, 2007.

[51] S. N. Gorb and E. V. Gorb. Ontogenesis of the attachment ability in the bug coreus marginatus (heteroptera, insecta). *Journal of Experimental Biology*, 207(17):2917–2924, 2004.

[52] E. Arzt, S. Gorb, and R. Spolenak. From micro to nano contacts in biological attachment devices. *PNAS*, 100(19):10603–10606, 2003.

[53] C.-Y. Hui, A. Jagota, L. Shen, A. Rajan, N. Glassmaker, and T. Tang. Design of bio-inspired fibrillar interfaces for contact and adhesion - theory and experiments. *Journal of Adhesion Science and Technology*, 21:1259–1280, 2007.

[54] M. Kamperman, E. Kroner, A. del Campo, R. M. McMeeking, and E. Arzt. Functional adhesive surfaces with "gecko" effect: The concept of contact splitting. *Advanced Engineering Materials*, 12(5):335–348, 2010.

[55] G. Carbone, E. Pierro, and S. N. Gorb. Origin of the superior adhesive performance of mushroom-shaped microstructured surfaces. *Soft Matter*, 7(12):5545–5552, 2011.

[56] B. N. J. Persson. On the mechanism of adhesion in biological systems. *Journal of Chemical Physics*, 118(16):7614–7621, 2003.

[57] A. M. Peattie, C. Majidi, A. Corder, and R. J. Full. Ancestrally high elastic modulus of gecko *Journal of The Royal Society Interface*, 4(17):1071–1076, 2007.

[58] H. Yao. *Mechanics of Robust and Releasable Adhesion in Biology*. PhD thesis, Universität Stuttgart, 2006.

[59] R. Spolenak, S. Gorb, and E. Arzt. Adhesion design maps for bio-inspired attachment systems. *Acta Biomaterialia*, 1(1):5–13, 2005.

[60] B. Bhushan. Adhesion of multi-level hierarchical attachment systems in gecko feet. *Journal of Adhesion Science and Technology*, 21:1213–1258, 2007.

[61] H. Yao and H. Gao. Mechanics of robust and releasable adhesion in biology: Bottom-up designed hierarchical structures of gecko. *Journal of the Mechanics and Physics of Solids*, 54(6):1120 – 1146, 2006.

[62] M. Sitti and R. S. Fearing. Synthetic gecko foot-hair micro/nano-structures as dry adhesives. *Journal of Adhesion Science and Technology*, 17:1055–1073, 2003.

[63] C. Y. Hui, A. Jagota, Y. Y. Lin, and E. J. Kramer. Constraints on microcontact printing imposed by stamp deformation. *Langmuir*, 18(4):1394–1407, 2002.

[64] A. Jagota and S. J. Bennison. Mechanics of adhesion through a fibrillar microstructure. *Integrative and Comparative Biology*, 42(6):1140–1145, 2002.

[65] N. J. Glassmaker; A. Jagota; C.-Y. Hui and J. Kim. Design of biomimetic fibrillar interfaces: 1. making contact. *Journal of The Royal Society Interface*, 1:35–48, 2004.

[66] H. Gao, X. Wang, H. Yao, S. Gorb, and E. Arzt. Mechanics of hierarchical adhesion structures of geckos. *Mechanics of Materials*, 37(2-3):275 – 285, 2005.

[67] N. J. Glassmaker, A. Jagota, and C.-Y. Hui. Adhesion enhancement in a biomimetic fibrillar interface. *Acta Biomaterialia*, 1(4):367 – 375, 2005.

[68] T. Tang, C.-Y. Hui, and N. J. Glassmaker. Can a fibrillar interface be stronger and tougher than a non-fibrillar one? *Journal of The Royal Society Interface*, 2(5):505–516, 2005.

[69] T. W. Kim and B. Bhushan. Optimization of biomimetic attachment system contacting with a rough surface. *The Journal of Vacuum Science and Technology A*, 25(4):1003–1012, 2007.

[70] C. Y. Hui, Y. Y. Lin, J. M. Baney, and A. Jagota. The accuracy of the geometric assumptions in the JKR (Johnson-Kendall-Roberts) theory of adhesion. *Journal of Adhesion Science and Technology*, 14:1297–1319(23), 18 July 2000.

[71] B. Schubert, C. Majidi, R. E. Groff, S. Baek, B. Bush, R. Maboudian, and R. S. Fearing. Towards friction and adhesion from high modulus microfiber arrays. *Journal of Adhesion Science and Technology*, 21:1297–1315, 2007.

[72] S. Liu, P. Zhang, X. Cheng, R. Malik, and Z. Tang. Optimal design of high-aspect-ratio micro/nano hierarchical structures mimicking gecko foot-hairs. *Advanced Science Letters*, 4(4-5):1546–1551, 2011.

[73] K. L. Johnson, K. Kendall, and A. D. Roberts. Surface energy and the contact of elastic solids. *Proceedings of the Royal Society of London. Series A, Mathematical and Physical Sciences*, 324(1558):301–313, 1971.

[74] D. G. Ellwood. *Mechanical Metallurgy*. New York, McGraw-Hill, 1961.

[75] H. G. Hahn. *Bruchmechanik*. Teubner Studienbücher: Mechanik, 1976.

[76] D. Gross and T. Seelig. *Bruchmechanik*. Springer, Berlin, 2006.

[77] M. Sitti. Oral communication, 2010. Department of Mechanical Engineering, Carnegie Mellon University.

[78] K. Y. Suh. Oral communication, 2010. School of Mechanical and Aerospace Engineering, Seoul National University, Seoul 151-744, Korea.

[79] F. Büsch. Oral communication, 2010. BASF.

[80] H. E. Jeong, S. H. Lee, P. Kim, and K. Y. Suh. Stretched polymer nanohairs by nanodrawing. *Nano Letters*, 6(7):1508–1513, 2006.

[81] J. Lee and R. S. Fearing. Contact self-cleaning of synthetic gecko adhesive from polymer microfibers. *Langmuir*, 24(19):10587–10591, 2008.

[82] M. T. Northen, C. Greiner, E. Arzt, and K. L. Turner. A gecko-inspired reversible adhesive. *Advanced Materials*, 20(20):3905–3909, 2008.

[83] L. Qu, L. Dai, M. Stone, Z. Xia, and Z. L. Wang. Carbon nanotube arrays with strong shear binding-on and easy normal lifting-off. *Science*, 322(5899):238–242, 2008.

[84] M. P. Murphy, B. Aksak, and M. Sitti. Gecko-inspired directional and controllable adhesion. *Small*, 5(2):170–175, 2009.

[85] H. Ko, Z. Zhang, J. C. Ho, K. Takei, R. Kapadia, Y.-L. Chueh, W. Cao, B. A. Cruden, and A. Javey. Flexible carbon-nanofiber connectors with anisotropic adhesion properties. *Small*, 6(1):22–26, 2010.

[86] Y. Zhang, C.-T. Lin, and S. Yang. Fabrication of Hierarchical Pillar Arrays from Thermoplastic and Photosensitive SU-8. *Small*, 6(6):768–775, 2010.

[87] M. K. Kwak, H. E. Jeong, W. G. Bae, H.-S. Jung, and K. Y. Suh. Adhesive microstructures: Anisotropic adhesion properties of triangular-tip-shaped micropillars. *Small*, 7(16):2266–2266, 2011.

[88] A. del Campo, C. Greiner, and E. Arzt. Contact shape controls adhesion of bioinspired fibrillar surfaces. *Langmuir*, 23(20):10235–10243, 2007.

[89] R. Spolenak, S. Gorb, H. Gao, and E. Arzt. Effects of contact shape on the scaling of biological attachments. *Proceedings of the Royal Society A: Mathematical, Physical and Engineering Science*, 461(2054):305–319, 2005.

[90] N. J. Glassmaker, A. Jagota, C.-Y. Hui, and J. Kim. Design of biomimetic fibrillar interfaces: 1. making contact. *Journal of The Royal Society Interface*, 1(1):23–33, 2004.

[91] M. Röhrig, M. Thiel, M. Worgull, and H. Hölscher. 3D direct laser writing of nano- and microstructured hierarchical gecko-mimicking surfaces. *Small*, 8(19):3009–3015, 2012.

[92] M. Deubel, G. von Freymann, M. Wegener, S. Pereira, K. Busch, and C. M. Soukoulis. Direct laser writing of three-dimensional photonic-crystal templates for telecommunications. *Nature Materials*, 3(7):444–447, 2004.

[93] C. Greiner, A. del Campo, and E. Arzt. Adhesion of bioinspired micropatterned surfaces: Effects of pillar radius, aspect ratio, and preload. *Langmuir*, 23(7):3495–3502, 2007.

[94] S. N. Gorb and M. Varenberg. Mushroom-shaped geometry of contact elements in biological adhesive systems. *Journal of Adhesion Science and Technology*, 21:1175–1183, 2007.

[95] N. Anscombe. Direct laser writing. *Nature Photonics*, 4:22–23, 2010.

[96] F. Marschall. Weiterentwicklung eines Gefriertrocknungsprozesses zur Herstellung von Mikro- and Nanostrukturen. Master's thesis, Karlsruher Institut für Technologie (KIT), Institut für Mikrostrukturtechnik (IMT), 2010.

[97] W. C. Oliver and G. M. Pharr. Measurement of hardness and elastic modulus by instrumented indentation: Advances in understanding and refinements to methodology. *Journal of Materials Research*, 19(01):3–20, 2004.

[98] T. S. Kustandi, V. D. Samper, D. K. Yi, W. S. Ng, P. Neuzil, and W. Sun. Self-assembled nanoparticles based fabrication of gecko foot-hair-inspired polymer nanofibers. *Advanced Functional Materials*, 17(13):2211–2218, 2007.

[99] A. Y. Y. Ho, H. Gao, Y. C. Lam, and I. Rodriguez. Controlled fabrication of multitiered three-dimensional nanostructures in porous alumina. *Advanced Functional Materials*, 18(14):2057–2063, 2008.

[100] A. Y. Y. Ho, L. P. Yeo, Y. C. Lam, and I. Rodriguez. Fabrication and analysis of gecko-inspired hierarchical polymer nanosetae. *ACS Nano*, 5(3):1897–1906, 2011.

[101] H.-J. Butt, B. Cappella, and M. Kappl. Force measurements with the atomic force microscope: Technique, interpretation and applications. *Surface Science Reports*, 59(1-6):1 – 152, 2005.

[102] L. H. Mak, M. Knoll, D. Weiner, A. Gorschluter, A. Schirmeisen, and H. Fuchs. Reproducible attachment of micrometer sized particles to atomic force microscopy cantilevers. *Review of Scientific Instruments*, 77(4):046104, 2006.

[103] E. Kroner, D. R. Paretkar, R. M. McMeeking, and E. Arzt. Adhesion of flat and structured pdms samples to spherical and flat probes: A comparative study. *The Journal of Adhesion*, 87(5):447–465, 2011.

[104] G. Castellanos, E. Arzt, and M. Kamperman. Effect of viscoelasticity on adhesion of bioinspired micropatterned epoxy surfaces. *Langmuir*, 27(12):7752–7759, 2011.

[105] D. Tabor. Surface forces and surface interactions. *Journal of Colloid and Interface Science*, 58(1):2–13, 1977.

[106] B. V. Derjaguin, V. M. Müller, and Y. P. Toporov. Effect of contact deformations on the adhesion of particles. *Journal of Colloid and Interface Science*, 53(2):314 – 326, 1975.

[107] D. Maugis. *Contact, Adhesion and Rupture of Elastic Solids*, volume 130 of *Springer Series in Solid-State Sciences*. Springer-Verlag, 2000.

[108] J. N. Israelachvili. *Intermolecular and surface forces*. Academic Press, 2. ed., 7. print. edition, 1998.

[109] V. L. Popov. *Kontaktmechanik und Reibung*. Springer Berlin Heidelberg, Berlin, Heidelberg, 2009.

[110] J. A. Greenwood and J. B. P. Williamson. Contact of nominally flat surfaces. *Proceedings of the Royal Society of London. Series A, Mathematical and Physical Sciences*, 295(1442):300–319, 1966.

[111] K. Autumn, A. Dittmore, D. Santos, M. Spenko, and M. Cutkosky. Frictional adhesion: A new angle on gecko attachment. *The Journal of Experimental Biology*, 209(18):3569–, 2006.

[112] D. Santos, A. Spenko, M.and Parness, S. Kim, and M. Cutkosky. Directional adhesion for climbing: theoretical and practical considerations. *Journal of Adhesion Science and Technology*, 21:1317–1341, 2007.

[113] J. Lee, C. Majidi, B. Schubert, and R. S. Fearing. Sliding-induced adhesion of stiff polymer microfibre arrays. I. Macroscale behaviour. *Journal of The Royal Society Interface*, 5(25):835–844, 2008.

[114] M. P. Murphy, B. Aksak, and M. Sitti. Adhesion and anisotropic friction enhancements of angled heterogeneous micro-fiber arrays with spherical and spatula tips. *Journal of Adhesion Science and Technology*, 21:1281–1296, 2007.

[115] T. Kim, H. E. Jeong, K. Y. Suh, and H. H. Lee. Stooped nanohairs: Geometry-controllable, unidirectional, reversible, and robust gecko-like dry adhesive. *Advanced Materials*, 21(22):2276–2281, 2009.

[116] S. Reddy, E. Arzt, and A. del Campo. Bioinspired surfaces with switchable adhesion. *Advanced Materials*, 19(22):3833–3837, 2007.

[117] A. K. Geim, S. V. Dubonos, I. V. Grigorieva, K. S. Novoselov, A. A. Zhukov, and S. Y. Shapoval. Microfabricated adhesive mimicking gecko foot-hair. *Nature Materials*, 2(7):461–463, 2003.

[118] S. Kim and M. Sitti. Biologically inspired polymer microfibers with spatulate tips as repeatable fibrillar adhesives. *Applied Physics Letters*, 89(26):261911, 2006.

[119] J. M. Karp and R. Langer. Materials science: Dry solution to a sticky problem. *Nature*, 477:42, 2011.

[120] B. N. J. Persson. Biological adhesion for locomotion on rough surfaces: Basic principles and a theorist's view. *Journal of Adhesion Science and Technology*, 21(12-13):1145–1173, 2007.

[121] C. Y. Hui, L. Shen, A. Jagota, and K. Autumn. Mechanics of anti-fouling or self-cleaning in gecko setae. In *Proceedings of the 29th Annual Meeting of The Adhesion Society*, pages 29–31, 2006.

[122] B. Sümer and M. Sitti. Rolling and spinning friction characterization of fine particles using lateral force microscopy based contact pushing. *Journal of Adhesion Science and Technology*, 22(5-6):481–506, 2008.

[123] C. Dominik and A. G. G. M. Tielens. Resistance to rolling in the adhesive contact of two elastic spheres. *Philosophical Magazine A*, 72(3):783–803, 1995.

[124] R. A. Potyrailo, H. Ghiradella, A. Vertiatchikh, K. Dovidenko, J. R. Cournoyer, and E. Olson. Morpho butterfly wing scales demonstrate highly selective vapour response. *Nature Photonics*, 1(2):123–128, 2007.

[125] A. D. Pris, Y. Utturkar, C. Surman, W. G. Morris, A. Vert, S. Zalyubovskiy, T. Deng, H. T. Ghiradella, and R. A. Potyrailo. Towards high-speed imaging of infrared photons with bio-inspired nanoarchitectures. *Nature Photonics*, 6:195–200, 2012.

[126] E. Kroner, J. S. Kaiser, S. C. L. Fischer, and E. Arzt. Bioinspired polymeric surface patterns for medical applications. *Journal of Applied Biomaterials and Functional Materials*, 10(3):287 – 292, 2012.

[127] H. E. Jeong and K. Y. Suh. Nanohairs and nanotubes: Efficient structural elements for gecko-inspired artificial dry adhesives. *Nano Today*, 4(4):335 – 346, 2009.

[128] L. P. Biró and J. P. Vigneron. Photonic nanoarchitectures in butterflies and beetles: valuable sources for bioinspiration. *Laser & Photonics Reviews*, 5(1):27–51, 2011.

[129] M. Worgull, A. Kolew, M. Heilig, M. Schneider, H. Dinglreiter, and B. Rapp. Hot embossing of high performance polymers. *Microsystem Technologies*, 17:585–592, 2011.

[130] P. Vella, S. Dimov, A. Kolew, E. Minev, K. Popov, F. Lacan, C. Griffiths, H. Hirshy, and S. Scholz. Bulk metallic glass based toolmaking process chain for micro- and nano- replication. In *Proceedings 4M Multi Material Micro Manufacture*, volume 9, pages 309–314, 2012.

[131] E.W. Becker, W. Ehrfeld, P. Hagmann, A. Maner, and D. Münchmeyer. Fabrication of microstructures with high aspect ratios and great structural heights by synchrotron radiation lithography, galvanoforming, and plastic moulding (liga process). *Microelectronic Engineering*, 4(1):35 – 56, 1986.

[132] M. Röhrig, M. Heilig, M. Schneider, A. Kolew, K. Kaiser, and M. Worgull M. Guttmann. Universelle Fixierung von Shim-Formeinsätzen für die Mikro- und Nanoreplikation auf Basis von Polymerfolien. *Galvanotechnik*, 101:1646–1653, 2010.

[133] G. Kumar, H. X. Tang, and J. Schroers. Nanomoulding with amorphous metals. *Nature*, 457(7231):868–872, 2009.

[134] W. Federle, W. J. P. Barnes, W. Baumgartner, P. Drechsler, and J. M. Smith. Wet but not slippery: boundary friction in tree frog adhesive toe pads. *Journal of The Royal Society Interface*, 3(10):689–697, 2006.

[135] J. Siegel, O. Lyutakov, V. Rybka, Z. Kolska, and V. Svorcik. Properties of gold nanostructures sputtered on glass. *Nanoscale Research Letters*, 6(1):96, 2011.

[136] S. Franssila. *Introduction to Micro Fabrication*. John Wiley & Sons, 2010.

[137] F. M. Smits. Measurement of sheet resistivities with the four-point probe. *Bell System Technical Journal*, May:711–718, 1958.

[138] L. Feng, Y. Zhang, J. Xi, Y. Zhu, N. Wang, F. Xia, and L. Jiang. Petal effect: A superhydrophobic state with high adhesive force. *Langmuir*, 24(8):4114–4119, 2008.

[139] X.-M. Li, D. Reinhoudt, and M. Crego-Calama. What do we need for a superhydrophobic surface? a review on the recent progress in the preparation of superhydrophobic surfaces. *Chemical Society Reviews*, 36:1350–1368, 2007.

[140] E. Bormashenko. Wetting transitions on biomimetic surfaces. *Philosophical Transactions of the Royal Society A: Mathematical,Physical and Engineering Sciences*, 368(1929):4695–4711, 2010.

[141] B. Bhushan, Y. C. Jung, and K. Koch. Micro-, nano- and hierarchical structures for superhydrophobicity, self-cleaning and low adhesion. *Philosophical Transactions of the Royal Society A: Mathematical,Physical and Engineering Sciences*, 367(1894):1631–1672, 2009.

[142] A. Parness, D. Soto, N. Esparza, N. Gravish, M. Wilkinson, K. Autumn, and M. Cutkosky. A microfabricated wedge-shaped adhesive array displaying gecko-like dynamic adhesion, directionality and long lifetime. *Journal of The Royal Society Interface*, 6(41):1223–1232, 2009.

[143] K. Liu, X. Yao, and L. Jiang. Recent developments in bio-inspired special wettability. *Chemical Society Reviews*, 39:3240–3255, 2010.

[144] J. Bekesi, J. Kaakkunen, W. Michaeli, F. Klaiber, M. Schoengart, J. Ihlemann, and P. Simon. Fast fabrication of super-hydrophobic surfaces on polypropylene by replication of short-pulse laser structured molds. *Applied Physics A: Materials Science & Processing*, 99:691–695, 2010.

[145] B. Farshchian, S. Park, J. Choi, A. Amirsadeghi, J. Lee, and S. Park. 3D nanomolding for lab-on-a-chip applications. *Lab on a Chip*, 12:4764–4771, 2012.

[146] N.-R. Chiou, C. Lu, J. Guan, L. J. Lee, and A. J. Epstein. Growth and alignment of polyaniline nanofibres with superhydrophobic, superhydrophilic and other properties. *Nature Nanotechnology*, 2(6):354–357, 2007.

[147] Y. Lee, S.-H. Park, K.-B. Kim, and J.-K. Lee. Fabrication of hierarchical structures on a polymer surface to mimic natural superhydrophobic surfaces. *Advanced Materials*, 19(17):2330–2335, 2007.

[148] L. Feng, S. Li, Y. Li, H. Li, L. Zhang, J. Zhai, Y. Song, B. Liu, L. Jiang, and D. Zhu. Super-hydrophobic surfaces: From natural to artificial. *Advanced Materials*, 14(24):1857–1860, 2002.

[149] Y. Liu and G. Li. A new method for producing lotus effect on a biomimetic shark skin. *Journal of Colloid and Interface Science*, 388(1):235 – 242, 2012.

[150] T. Nishino, M. Meguro, K. Nakamae, M. Matsushita, and Y. Ueda. The lowest surface free energy based on CF_3 alignment. *Langmuir*, 15(13):4321–4323, 1999.

[151] N. Zhao, J. Xu, Q. Xie, L. Weng, X. Guo, X. Zhang, and L. Shi. Fabrication of biomimetic superhydrophobic coating with a micro-nano-binary structure. *Macromolecular Rapid Communications*, 26(13):1075–1080, 2005.

[152] M. Michaeli, M. Schoengart, F. Klaiber, and S. Beckemper. Variothermal injection molding of superhydrophobic surfaces. In S. Dimov B. Fillon, C. Khan-Malek, editor, *Proceedings of the 7th International Conference on Multi-Material Micro Manufacture*. Reseach Publishing, 2010.

[153] T.-S. Wong, S. H. Kang, S. K. Y. Tang, E. J. Smythe, B. D. Hatton, A. Grinthal, and J. Aizenberg. Bioinspired self-repairing slippery surfaces with pressure-stable omniphobicity. *Nature*, 477(7365):443–447, 2011.

[154] H. Reiss. Warum gibt es Benetzung? *Physik in unserer Zeit*, 23(5):204–212, 1992.

[155] T. Young. An essay on the cohesion of fluids. *Philosophical Transactions of the Royal Society of London*, 95:65–87, 1805.

[156] R. N. Wenzel. Resistance of solid surfaces to wetting by water. *Industrial & Engineering Chemistry*, 28(8):988–994, 1936.

[157] D. Quéré. Surface chemistry: Fakir droplets. *Nature Materials*, 1(1):14–15, 2002.

[158] D. Quéré. Non-sticking drops. *Reports on Progress in Physics*, 68(11):2495, 2005.

[159] S. Baxter A. B. D. Cassie. Wettability of porous surfaces. *Transactions of the Faraday Society*, 40:546, 1944.

[160] W. Chen, A. Y. Fadeev, M. C. Hsieh, D. Oener, J. Youngblood, and T. J. McCarthy. Ultrahydrophobic and ultralyophobic surfaces: Some comments and examples. *Langmuir*, 15(10):3395–3399, 1999.

[161] D. Quéré, M.-J. Azzopardi, and L. Delattre. Drops at rest on a tilted plane. *Langmuir*, 14(8):2213–2216, 1998.

[162] Z. Yoshimitsu, A. Nakajima, T. Watanabe, and K. Hashimoto. Effects of surface structure on the hydrophobicity and sliding behavior of water droplets. *Langmuir*, 18(15):5818–5822, 2002.

[163] D. Rosato and M. Rosato. *Injection Molding Handbook*, volume 3. ed. / XXXI, 1457 S. : Ill. Kluwer Academic, 2000.

[164] A. Pépin Y. Chen. *Nanoscience - Nanotechnologies and Nanophysics*, chapter Emerging Nanolithographic Methods, pages 157–175. Springer, 2007. Claire Dupas, Philippe Houdy, Marcel Lahmani.

[165] T. S. Kustandi, V. D. Samper, W. S. Ng, A. S. Chong, and H. Gao. Fabrication of a gecko-like hierarchical fibril array using a bonded porous alumina template. *Journal of Micromechanics and Microengineering*, 17(10):N75, 2007.

[166] H. Tan, A. Gilbertson, and S. Y. Chou. Roller nanoimprint lithography. *Journal of Vacuum Science Technology B: Microelectronics and Nanometer Structures*, 16(6):3926 –3928, 1998.

[167] S. H. Ng and Z. F. Wang. Hot roller embossing for microfluidics: process and challenges. *Microsystem Technologies*, 15:1149–1156, 2009.

[168] T. Onda, S. Shibuichi, N. Satoh, and K. Tsujii. Super-water-repellent fractal surfaces. *Langmuir*, 12(9):2125–2127, 1996.

[169] J. Bico, C. Marzolin, and D. Quéré. Pearl drops. *Europhysics Letters*, 47(2):220, 1999.

[170] M. Callies and D. Quéré. On water repellency. *Soft Matter*, 1(1):55–61, 2005.

[171] B. Bhushan. Personal communication.

[172] M. Nosonovsky and B. Bhushan. Energy transitions in superhydrophobicity: low adhesion, easy flow and bouncing. *Journal of Physics: Condensed Matter*, 20(39):395005, 2008.

[173] R. J. Daniello, N. E. Waterhouse, and J. P. Rothstein. Drag reduction in turbulent flows over superhydrophobic surfaces. *Physics of Fluids*, 21(8):085103, 2009.

[174] E. Pierce, F. J. Carmona, and A. Amirfazli. Understanding of sliding and contact angle results in tilted plate experiments. *Colloids and Surfaces A: Physicochemical and Engineering Aspects*, 323(1-3):73 – 82, 2008.

[175] B. Bhushan and E. K. Her. Fabrication of superhydrophobic surfaces with high and low adhesion inspired from rose petal. *Langmuir*, 26(11):8207–8217, 2010.

[176] P.-G. de Gennes, F. Brochard-Wyard, and D. Quéré. *Capillarity and wetting phenomena: drops, bubbles, pearls, waves*. Springer Science+Business Media, New-York, 2004.

[177] H. F. Bohn and W. Federle. Insect aquaplaning: Nepenthes pitcher plants capture prey with the peristome, a fully wettable water-lubricated anisotropic surface. *PNAS*, 101(39):14138–14143, 2004.

[178] B. Pokroy, A. K. Epstein, M. C. M. Persson-Gulda, and J. Aizenberg. Fabrication of bioinspired actuated nanostructures with arbitrary geometry and stiffness. *Advanced Materials*, 21(4):463–469, 2009.

[179] C. Ishino, M. Reyssat, E. Reyssat, K. Okumura, and D. Quéré. Wicking within forests of micropillars. *Europhysics Letters*, 79(5):56005, 2007.

[180] J. Kim and C.-J. Kim. Nanostructured surfaces for dramatic reduction of flow resistance in droplet-based microfluidics. In *The Fifteenth IEEE International Conference on Micro Electro Mechanical Systems*, pages 479 –482, 2002.

[181] J.Ou, B. Perot, and J. P. Rothstein. Laminar drag reduction in microchannels using ultrahydrophobic surfaces. *Physics of Fluids*, 16(12):4635–4643, 2004.

[182] F. Feuillebois, M. Z. Bazant, and O. I. Vinogradova. Effective slip over superhydrophobic surfaces in thin channels. *Physical Review Letters*, 102:026001, 2009.

[183] C. Lee and C.-J. Kim. Maximizing the giant liquid slip on superhydrophobic microstructures by nanostructuring their sidewalls. *Langmuir*, 25(21):12812–12818, 2009.

[184] P. Ditsche-Kuru, E. S. Schneider, J.-E. Melskotte, M. Brede, A. Leder, and W. Barthlott. Superhydrophobic surfaces of the water bug notonecta glauca: a model for friction reduction and air retention. *Beilstein Journal of Nanotechnology*, 2:137–144, 2011.

[185] P. S. Nobel. *Physicochemical and Environmental Plant Physiology*. Amsterdam: Elsevier Academic Press, 2005.

[186] O. K. Rediniotis A. K. Balasubramanian, A. C. Miller. Microstructured hydrophobic skin for hydrodynamic drag reduction. *The American Institute of Aeronautics and Astronautics Journal*, 42:411, 2004.

[187] C. Henoch, T. N. Krupenkin, P. Kolodner, J. A. Taylor, M. S. Hodes, A. M. Lyons, and K. Breuer. Turbulent drag reduction using superhydrophobic surfaces. *3rd AIAA Flow Control Conference*, 5 - 8 June 2006, San Francisco, California:3192, 2006.

[188] G. S. Kino and T. R. Corle. Confocal scanning optical microscopy. *Physics Today*, 42(9):55–62, 1989.

[189] T. H. Ribeiro, J. Rubio, and R. W. Smith. A dried hydrophobic aquaphyte as an oil filter for oil/water emulsions. *Spill Science & Technology Bulletin*, 8(5-6):483 – 489, 2003.

[190] M.O. Adebajo, R. L. Frost, J. T. Kloprogge, O. Carmody, and S. Kokot. Porous materials for oil spill cleanup: A review of synthesis and absorbing properties. *Journal of Porous Materials*, 10:159–170, 2003.

[191] C. Robertson and C. Krauss. Gulf spill is the largest of its kind, scientists say. *New York Times*, August 2, 2010.

[192] National Commission on the BP Deepwater Horizon Oil Spill and Offshore Drilling. Deep water - the gulf oil desaster and the future of offshore drilling. *Report to the President*, 2011.

[193] J. T. Korhonen, M. Kettunen, R. H. A. Ras, and O. Ikkala. Hydrophobic nanocellulose aerogels as floating, sustainable, reusable, and recyclable oil absorbents. *ACS Applied Materials & Interfaces*, 3(6):1813–1816, 2011.

[194] X. Chen, L. Hong, Y. Xu, and Z. W. Ong. Ceramic pore channels with inducted carbon nanotubes for removing oil from water. *ACS Applied Materials & Interfaces*, 4(4):1909–1918, 2012.

[195] A. V. Rao, N. D. Hegde, and H. Hirashima. Absorption and desorption of organic liquids in elastic superhydrophobic silica aerogels. *Journal of Colloid and Interface Science*, 305(1):124 – 132, 2007.

[196] Q. Wen, J. Di, L. Jiang, J. Yu, and R. Xu. Zeolite-coated mesh film for efficient oil-water separation. *Chemical Science*, 4(2):591–595, 2013.

[197] J. G. Reynolds, P. R. Coronado, and L. W. Hrubesh. Hydrophobic aerogels for oil-spill clean up - synthesis and characterization. *Journal of Non-Crystalline Solids*, 292:127 – 137, 2001.

[198] Q. F. Wei, R. R. Mather, A. F. Fotheringham, and R. D. Yang. Evaluation of nonwoven polypropylene oil sorbents in marine oil-spill recovery. *Marine Pollution Bulletin*, 46(6):780–783, 2003.

[199] F. M. Fowkes. Attractive forces at interfaces. *Industrial & Engineering Chemistry*, 56(12):40–52, 1964.

[200] D. K. Owens and R. C. Wendt. Estimation of the surface free energy of polymers. *Journal of Applied Polymer Science*, 13(8):1741–1747, 1969.

[201] M. J. Schwuger, G. H. Findenegg, and F.-H. Haegel. *Lehrbuch der Grenzflächenchemie*. Wiley-Vch, 1996.

[202] J. J. Jasper. The surface tension of pure liquid compounds. *Journal of Physical and Chemical Reference Data*, 1(4):841–1010, 1972.

[203] Chen Jie-Rong and T. Wakida. Studies on the surface free energy and surface structure of ptfe film treated with low temperature plasma. *Journal of Applied Polymer Science*, 63(13):1733–1739, 1997.

[204] http://www.reflexite.com.

[205] L.-T. Jiang, T.-C. Huang, C.-Y. Chang, J.-R. Ciou, S.-Y. Yang, and P.-H. Huang. Direct fabrication of rigid microstructures on a metallic roller using a dry film resist. *Journal of Micromechanics and Microengineering*, 18(1):015004, 2008.

[206] H. Peisker. Physical properties of materials and liquids in insect adhesive systems probed by AFM. In *4th International Workshop on Advanced Atomic Force Microscopy Techniques, Karlsruhe, March 4 -5*, 2013.

[207] United Nations Environment Programme (UNEP) and OSPAR Commission. Marine litter, 2009.

Schriften des Instituts für Mikrostrukturtechnik
am Karlsruher Institut für Technologie (KIT)

ISSN 1869-5183

Herausgeber: Institut für Mikrostrukturtechnik

Die Bände sind unter www.ksp.kit.edu als PDF frei verfügbar
oder als Druckausgabe zu bestellen.

Band 1 Georg Obermaier
Research-to-Business Beziehungen: Technologietransfer durch Kommunikation von Werten (Barrieren, Erfolgsfaktoren und Strategien). 2009
ISBN 978-3-86644-448-5

Band 2 Thomas Grund
Entwicklung von Kunststoff-Mikroventilen im Batch-Verfahren. 2010
ISBN 978-3-86644-496-6

Band 3 Sven Schüle
Modular adaptive mikrooptische Systeme in Kombination mit Mikroaktoren. 2010
ISBN 978-3-86644-529-1

Band 4 Markus Simon
Röntgenlinsen mit großer Apertur. 2010
ISBN 978-3-86644-530-7

Band 5 K. Phillip Schierjott
Miniaturisierte Kapillarelektrophorese zur kontinuierlichen Überwachung von Kationen und Anionen in Prozessströmen. 2010
ISBN 978-3-86644-523-9

Band 6 Stephanie Kißling
Chemische und elektrochemische Methoden zur Oberflächenbearbeitung von galvanogeformten Nickel-Mikrostrukturen. 2010
ISBN 978-3-86644-548-2

Schriften des Instituts für Mikrostrukturtechnik
am Karlsruher Institut für Technologie (KIT)

ISSN 1869-5183

Band 7 Friederike J. Gruhl
Oberflächenmodifikation von Surface Acoustic Wave (SAW) Biosensoren für biomedizinische Anwendungen. 2010
ISBN 978-3-86644-543-7

Band 8 Laura Zimmermann
Dreidimensional nanostrukturierte und superhydrophobe mikrofluidische Systeme zur Tröpfchengenerierung und -handhabung. 2011
ISBN 978-3-86644-634-2

Band 9 Martina Reinhardt
Funktionalisierte, polymere Mikrostrukturen für die dreidimensionale Zellkultur. 2011
ISBN 978-3-86644-616-8

Band 10 Mauno Schelb
Integrierte Sensoren mit photonischen Kristallen auf Polymerbasis. 2012
ISBN 978-3-86644-813-1

Band 11 Daniel Auernhammer
Integrierte Lagesensorik für ein adaptives mikrooptisches Ablenksystem. 2012
ISBN 978-3-86644-829-2

Band 12 Nils Z. Danckwardt
Pumpfreier Magnetpartikeltransport in einem Mikroreaktionssystem: Konzeption, Simulation und Machbarkeitsnachweis. 2012
ISBN 978-3-86644-846-9

Band 13 Alexander Kolew
Heißprägen von Verbundfolien für mikrofluidische Anwendungen. 2012
ISBN 978-3-86644-888-9

Schriften des Instituts für Mikrostrukturtechnik
am Karlsruher Institut für Technologie (KIT)

ISSN 1869-5183

Band 14 Marko Brammer
Modulare Optoelektronische Mikrofluidische Backplane. 2012
ISBN 978-3-86644-920-6

Band 15 Christiane Neumann
Entwicklung einer Plattform zur individuellen Ansteuerung von Mikroventilen und Aktoren auf der Grundlage eines Phasenübergangs zum Einsatz in der Mikrofluidik. 2013
ISBN 978-3-86644-975-6

Band 16 Julian Hartbaum
Magnetisches Nanoaktorsystem. 2013
ISBN 978-3-86644-981-7

Band 17 Johannes Kenntner
Herstellung von Gitterstrukturen mit Aspektverhältnis 100 für die Phasenkontrastbildgebung in einem Talbot-Interferometer. 2013
ISBN 978-3-7315-0016-2

Band 18 Kristina Kreppenhofer
Modular Biomicrofluidics - Mikrofluidikchips im Baukastensystem für Anwendungen aus der Zellbiologie. 2013
ISBN 978-3-7315-0036-0

Band 19 Ansgar Waldbaur
Entwicklung eines maskenlosen Fotolithographiesystems zum Einsatz im Rapid Prototyping in der Mikrofluidik und zur gezielten Oberflächenfunktionalisierung. 2013
ISBN 978-3-7315-0119-0

Band 20 Christof Megnin
Formgedächtnis-Mikroventile für eine fluidische Plattform. 2013
ISBN 978-3-7315-0121-3

Schriften des Instituts für Mikrostrukturtechnik
am Karlsruher Institut für Technologie (KIT)

ISSN 1869-5183

Band 21 Srinivasa Reddy Yeduru
Development of Microactuators Based on the Magnetic Shape Memory Effect. 2013
ISBN 978-3-7315-0125-1

Band 22 Michael Röhrig
Fabrication and Analysis of Bio-Inspired Smart Surfaces. 2014
ISBN 978-3-7315-0163-3